FOOD SCIENCE AND TECHNOLOGY

SENN'S WAR TIME COOKING GUIDE

FOOD SCIENCE AND TECHNOLOGY

Additional books and e-books in this series can be found on Nova's website under the Series tab.

AMERICAN POLITICAL, ECONOMIC, AND SECURITY ISSUES

Additional books and e-books in this series can be found on Nova's website under the Series tab.

FOOD SCIENCE AND TECHNOLOGY

SENN'S WAR TIME COOKING GUIDE

CHARLES HERMAN SENN

snova
New York

Library of Congress Cataloging-in-Publication Data

ISBN: 978-1-53615-023-0

Published by Nova Science Publishers, Inc. † New York

Contents

PREFACE*

This book, written by Charles Herman Senn (1864-1934) and originally published in 1915, brings the old recipes and techniques of cooking from during the war. It provides a unique window into peoples' lives and how they survived during the war. The book is broken down into the following sections:

- Recipes for Soups
- Fish Preparations
- Meat Dishes
- Vegetables
- Breakfast and Supper Dishes
- Sweets, Puddings
- Cakes and Pastries
- Invalid Dishes
- Sauces
- Management of Gas Stoves
- Inexpensive Family Meals

---

* This is an edited, reformatted and augmented version of "Senn's War Time Cooking Guide" by Charles Herman Senn and was previously published by The Food & Cookery Publishing Co., Westminister, S.W., London in 1915.

- Some Economy Hints (substitutes, for example, the can be used for Meat and Fish which were scare, how to avoid waste in food scraps and how to avoid waste in cooking, etc.).
- Index

Chapter 1

WAR TIME AND FOOD ECONOMY

However much we, as civilized individuals, may deplore the present disastrous European War, with its burden of militarism, brutal mechanical slaying devices, and terrible loss of life, there is no doubt that it is also the cause of many a "silver lining," not the least of which is a definite clarion call to return to a simpler mode of living.

On all hands one hears of serious cases of unemployment, reduction of income, even total loss, financial stresses of every sort and kind; and simultaneously, taxes necessarily increase, and food prices rise steadily higher. Who is sufficient for all these things? Again, there is but the one solution—return to a simpler life.

A thorough acquaintance with really nutritive foods, and a fixed determination to cut out from the daily dietary all those dishes which, though custom and tradition have prescribed their use, we know at the bottom of our hearts are not really necessary, will prove two most valuable assets to housewives in war time, if they will but give them a trial.

The present prohibitive prices of meat and fish have already caused many people to partially, and in many cases wholly, abandon their use, if not permanently, at any rate during the present crisis. The economical housewife can reduce her expenses considerably without becoming a faddist or a whole-hearted vegetarian. By using less meat and the cheaper kinds of fish, and substituting for them such cheap dishes as recommended in this booklet, an appreciable economy can be effected.

Now we eat meat and fish primarily for their proteid, or flesh-forming qualities, but everybody knows that there are many adequate substitutes for flesh foods upon the market, less costly in many cases, and every bit as nourishing. Cheese, peas, beans, lentils, commonly known as pulse foods—wholemeal flour, maize, semolina, oatmeal, eggs, and a very neglected article of food, nuts and nutmeat, all contain proteid, or flesh-forming properties, equally with meat and fish. One glance at the average prices of these substitutes compared with those now obtaining at the butcher's and fishmonger's shops, is quite enough to convince any intelligent housewife that she would be several pennies in pocket by adopting such substitutes during the coming months.

People who habitually consume a quantity of butcher's meat, or fish, are really living upon vegetable foods at second-hand. It will be found far more satisfactory to go to the vegetable world direct, besides being much cheaper. Most nuts (with the exception of chestnuts and coconut) are equal to meat and fish in proteid value, and are much superior as regards fat. All nuts must be well chewed. If the teeth are not good, the nuts may be ground by means of a nut mill, when they will be found quite easy to masticate. The quantity of nuts required (in place of meat) by a person of average weight may be anything between 3 and 8 oz. per day. Sedentary workers usually require less than active folk. Take, for instance, the following table of flesh substitutes, and note their food value as regards proteid, and their cheapness in price, as compared with the present prices of beef and fish.

Flesh-Forming Foods

Food.	Amount of Proteid in 100 parts.	Cost (War Prices).
Wholemeal	10·4	2d. per lb.
Oatmeal	16·1	2½d. ,,
Peas	22·4	3½d. ,,
Haricots	23·0	2½d. ,,
Lentils	25·0	3¼d. ,,
Cheese	29·2	9½d. ,,
Eggs	14·0	1½d. each.
Peanuts	24·5	5d. per lb.
As compared with :		
Beef	19·5	10d. to 1s. per lb.
Pork	20·8	9d. ,, 11d. ,,
Mutton	19·0	7d. ,, 9d. ,,
Flounder	14·0	7d. per lb.
Herring	10·9	1/3 to 1/6 per doz.

Carbohydrates, viz., starch, sugar, and fats, are necessary as proteid foods in a well-balanced dietary. These give heat and energy to the system, and are eaten, roughly speaking, in the proportion of three times as much as the proteid foods. To take the place of sugar, which has risen rapidly in price since war broke out, and is likely to remain high for some time, dried fruits, such as dates, raisins, figs, etc., will be found equally acceptable, and far more nourishing.

The chief sources of carbohydrate foods are in grains or cereals. Here, too, economy may be effected by a wise selection. Finely ground wholemeal flour costs rather more than white flour, but bread, biscuits, etc., made with the wholemeal, satisfy hunger more quickly, and supply more nourishment in the long run. Brown flour is also richer than white flour in bone and blood forming material. Again, semolina, macaroni, maize, and unpolished or "natural" rice, supply a far better food value than sago, tapioca, and cornflour, even if they work out at a little more in price. Like the flesh substitutes mentioned earlier, they are not really dearer in the long run, as less of these foods is required, compared with carbohydrates of poorer food value.

Fat is an important element in diet, and hitherto has been one of the most expensive. Fresh dairy butter and cream are beyond the reach of a good many people during the winter months, especially in war time, and their place is usually taken by lard, and dripping, and possibly margarine. These are not considered equal in food value to nut butters and other vegetable fats now easily to be obtained at most of our grocers' shops. The richness of nuts in *fat* as well as proteid has already been mentioned, and may be seen more clearly by a glance at the following table of carbohydrate foods. Their cheapness as compared with animal fats is also well known.

CARBOHYDRATE FOODS

Food.	Amount in 100 parts.		Cost (War Prices).
	Starch.	Fat.	
Macaroni	73·8	·9	5d. per lb.
Maize	66·5	5·0	3d. ,,
Figs	60·5	·9	5d. ,,
Dates	54·0	·2	4d. ,,
Peanuts	11·7	50·0	5d. ,,
Sago	83·0	...	3d. ,,
Tapioca	83·0	...	3d. ,,
Nut margarine	...	80·0	6d. to 8d. per lb.
Dairy butter	...	87·7	1/6 per lb.

The last class of foods consists of Bonemaking and blood-purifying ingredients, and are equally as important as the other two classes, although many people are inclined to estimate them too lightly. A liberal daily supply of foods which contain these properties is absolutely imperative to the maintenance of good bodily health. All root, tuber, stalk, and leaf vegetables, fresh fruits, and plenty of wholesome water, drunk daily in the proportion of about 1 to 2 quarts, will be found to supply the necessary elements. Whenever possible, the vegetables and fruits should be eaten raw, otherwise a great deal of goodness is lost in the cooking.

The water in which vegetables are cooked receives this goodness, and is too often relegated, in consequence, to the kitchen sink. To get the most food value out of cooked vegetables, they should be steamed, or baked in a closed vessel, with very little water, or made into soups and stews, so that all the goodness is *conserved,* or kept in. Fresh fruit, or stewed, used moderately and with discrimination, is not an expensive item, if care is observed to purchase only those kinds really in season, and therefore plentiful and cheap. The special food value of fruits and vegetables lies in their valuable acids and salts. They also contain sugar in its purest form, ready for immediate absorption into the system.

Many people are afraid to drop meat and fish out of their dietary, in case they should starve themselves.

This fear, of course, arises from ignorance of the food values of the flesh substitutes enumerated above. Indeed, it is easier to overeat on vegetarian fare than on flesh foods. Besides, it is not the quantity of food eaten that nourishes us, but the amount digested.

A *wise* use of food scraps is of the utmost importance at this critical time. Stale bread, dripping, bones, bacon rinds, vegetable parings, all have their uses.

Stale bread may be employed in sweet and savoury dishes, and as stuffings for meat and fish. Bones, bacon rinds, and a few soup vegetables, if stewed slowly in water for an hour or so, will lay the foundation of good economical stock, to be used in the making of soups and gravies.

Meals, too, need revision nowadays. These have been enlarged or added to very considerably during the last few years, and it is high time that we harked back to a simpler mode of living as well as eating. Abernethy's advice to "live on sixpence a day and earn it" would still work miracles if it were more universally applied. To drop out one meal during the day, at any rate whilst the war lasts, would hurt nobody, and would be gratefully recognised by the harassed housekeeper as at least one method of helping to make both ends meet. The remaining meals will be all the better enjoyed and digested, especially if proper attention is paid to thorough mastication, and: liquids are drunk after, and not whilst eating.

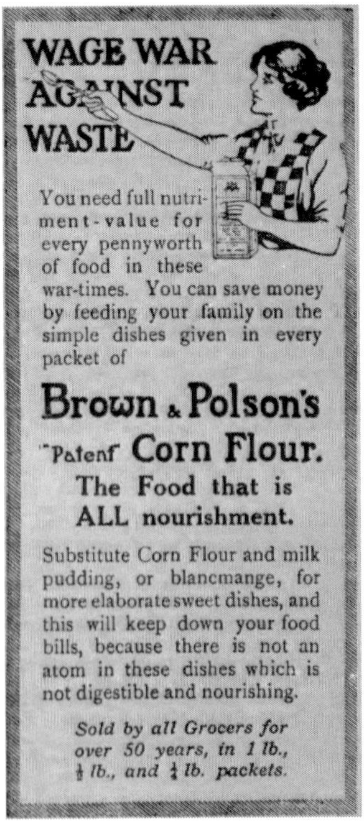

Chapter 2

RECIPES FOR SOUPS

BONE SOUP

2d. worth of bones, chopped small, 2 oz. dripping, 2 large onions, peeled and cut into slices, 2 carrots scraped and sliced, 1 turnip peeled and sliced, ½ pint split peas or lentils, 1 teaspoonful mixed herbs, salt and pepper to taste. Average cost, 5d.

Put the bones and dripping into a saucepan, and fry a little over the fire to brown slightly, then add 2 quarts of cold water and bring slowly to the boil. Skim, and add the vegetables and peas or lentils (previously soaked). Stir till it boils, then simmer for 2 hours or longer.

Remove the bones, add the herbs, and season to taste, then serve. The bones may be used a second time with fresh vegetables.

VEGETABLE SOUP (NOURISHING)

1 leek, 1 onion, ¼ small cabbage, 2 oz. dripping or butter, 1½ lb. potatoes, 1 pint milk, 1 heaped-up tablespoonful crushed tapioca or sago, salt and pepper. Average cost, 6d.

Wash and trim the leek and the cabbage, peel the onion, cut all into small slices. Wash and peel and slice the potatoes. Put all the vegetables in a large saucepan with the dripping or butter, and stir over the fire for some minutes. Moisten with 3 or 4 pints of water, and leave to simmer for about an hour; season to taste with pepper and salt, and rub through a colander or wire sieve. Boil up, and add the milk to the soup. Let all come to the boil, skim, then stir in the tapioca or sago, and simmer for another 10 minutes. Stir occasionally. A teaspoonful of celery seed, or a few outside leaves of fresh celery washed and cut into shreds, can also be added, if liked.

Fish Soup

1 lb. white fish (haddock, hake or cod), ½ lb. potatoes, 1 onion, 1 gill of milk, 1 quart of water, 1 oz. dripping or butter, seasoning. Average cost, 8d.

Clean the fish, and cut up into neat pieces. Wash, peel and slice the potatoes, peel and slice onion. Melt the fat in a saucepan. When hot, put in the vegetables, and toss lightly over the fire for a few minutes. Then add the fish and water, and cook slowly till tender. Rub the soup through a colander or sieve, return to the saucepan, add the milk and seasoning. Boil up again, then serve. A little chopped parsley may be added before serving.

Beef Soup

1 lb. of shin of beef, 1 onion, 1 oz. dripping, 1 turnip, ¼ lb. rice, 1 carrot, 2 quarts of water, salt and pepper. Average cost, 7d.

Cut the meat into small pieces, peel the onion and cut up small. Melt the dripping in a saucepan, when hot, put in the onion and fry a little; now add the beef and fry a nice brown, then add the water. Let it come to the boil, take off the scum, and add the vegetables, cleaned and cut up small, and the rice. Let the soup simmer gently for about 2 hours, stir and skim occasionally, season with pepper and salt, and serve hot. If liked, the meat may be cooked whole in the soup as a joint, and eaten as a second course.

Pea Soup

½ pint split peas, 1 stick of celery, a small carrot, turnip, and 1 small onion (peeled thin), 1 oz. dripping, ½ pint skim milk, 1 quart of stock or water, salt and pepper. Average cost, 4d, if made with water.

Soak the peas overnight. Melt the dripping in a saucepan, put in the peas and sliced vegetables. Fry lightly for ten minutes. Add the stock or water, and boil gently for about 2 hours. When the peas are quite soft, rub them through a sieve or colander.

Pour the soup into the saucepan and boil up, add salt and pepper and the milk, and serve with powdered mint. The stock from salt meat, such as a ham bone, can be used for pulse soups.

Lentil Soup

This is prepared the same as pea soup by merely substituting for the peas ½ pint of red lentils.

MUTTON BROTH

1 lb. scrag end of neck of mutton, 1 small carrot, small turnip, 1 small onion, 1 quart of water, 2 oz. pearl barley, 1 teaspoonful of salt, ½ teaspoonful of parsley. Average cost, 8d.

Trim off some of the fat from the mutton, and cut it into small pieces, and put it in the saucepan with a quart of cold water, and a teaspoonful of salt. Boil for an hour, then skim well. Add the prepared vegetables, previously cut into dice, and pearl barley, which should be well washed, and leave to simmer for another hour. Season with salt and pepper, and add the chopped parsley before serving. If liked, the meat may be cooked whole in the soup, and dished separately.

POTATO SOUP

1 lb. raw potatoes, 1 quart stock or water, 1 onion, ½ pint milk, 1 oz. butter or dripping, salt and pepper, a little parsley if handy, 1 oz. fine sago. Average cost, 3½d. if made with water.

Wash, peel and slice the potatoes and onion, melt the fat in a saucepan, put in the vegetables when hot, cook over the fire for a few minutes, then add the stock or water. Cook slowly till tender, then rub through a colander or sieve and return to the saucepan, season with pepper and salt to taste. Add the milk and sago, boil up again, and serve with a little finely chopped parsley on top.

MACARONI SOUP

3 oz. macaroni (small), 1½ pints stock, 2 oz. of grated cheese, salt and pepper. Average cost,

Blanch the macaroni in salted water, drain, cool, and cut it into short pieces. Cook it for about half an hour in the stock, and add about half a pint of water. Season with salt and pepper. Skim the soup as it boils up first. Serve with a little grated cheese.

French Rice Soup

2½ oz. rice, 2 oz. butter, 1 onion, 1 clove, 1½ pints stock, pint milk, salt, pepper, nutmeg. Average cost, 6d.

Wash and drain the rice, put it in a saucepan with the butter, previously melted, stir over the fire for a few minutes, then add the stock. Boil up, skim, add the onion, peeled and stuck with a clove, also the milk. Simmer gently for about 35 minutes. Season to taste with salt, pepper, and a grate of nutmeg. Stir occasionally whilst cooking to prevent burning. Remove the onion before serving.

Haricot Bean Soup

½ pint of haricot beans, 2 small potatoes, 1 onion, 1 turnip, 1 quart of stock or water, ½ pint skim milk, seasoning. Cost, 5d. if made with water.

Wash the beans and soak them in cold water overnight. Put them in a stewpan with the water or stock, let it come to the boil, and skim. Wash the vegetables, peel and slice them, and add to the soup. Boil gently for 2 hours, or until tender. Rub all through a sieve or colander, add the milk, return the soup to the stewpan, and boil up again. Season with salt and pepper, and serve.

Winter Soup

1 teacupful lentils, ½ teacupful oatmeal, 1 teacupful mashed potato, I breakfast cupful, in all, of chopped onion, carrot, turnip, salt and pepper, 3½ pints water. Cost, 4d.

Soak the lentils overnight, drain, put into a saucepan with the water, cook one hour, add the vegetables, salt and pepper, cook half an hour longer, sprinkle in the oatmeal, and add the mashed potato. Stir frequently, simmer gently until the lentils are quite soft, and serve. When coarse oatmeal is used, the soup must simmer half an hour after adding it.

Note

This soup may also be mashed through a sieve or colander before serving, and a little milk added to enrich it.

Cabbage Soup

1 cabbage, 1 oz. dripping, 1 pint skim milk, 2 pints water, 1 oz. fine sago, seasoning. Average cost, 4d.

Wash and clean the cabbage, and cut it up small. Melt the dripping in a saucepan, and toss the cabbage lightly in it for a few minutes. Cover with the water, and allow to simmer for of an hour. Then add the sago, milk, seasoning, and boil up again until the former is transparent. Serve with sippets of fried or toasted bread.

Chapter 3

Fish Preparations

Boiled Fish

Clean the fish, scale if necessary, trim, etc., and wash it in cold water. Put a saucepan or fish-kettle on the fire with sufficient water to well cover the fish; when the water is warm put in the fish. Season with salt and vinegar and let it simmer gently, allowing from 6 to 10 minutes to each pound, according to the thickness, with a few minutes over; skim occasionally. When the flesh has a creamy appearance and conies away easily from the bones, it is sufficiently boiled. Take it out carefully with a fish-slice, drain well, and place on a hot dish.

Fish Fried in Batter

1 plaice or haddock, or any other white flat-fish, 4 oz. flour, a gill of tepid water, 1 tablespoonful of oil or butter, salt, and fat for frying. Average cost, 4d. to 7d.

Wash the fish, dry and fillet it, and cut into convenient pieces. Prepare the batter as follows: Put the flour into a basin, add a pinch of salt, stir in the oil and sufficient tepid water to form a smooth batter free from lumps, allow it to stand for half an hour or longer. Heat the fat, dip each piece of fish into the batter and drop into the fat (smoking hot). Fry a light brown, drain on paper, and serve on a hot plate or dish.

The fish may be cut up without filleting and be fried in this way.

Fried Herrings (Scotch Fashion)

Average cost, 1d. 1½d. to for each herring.

Remove the heads from the required number of herrings. Split them open, and take out the backbones. Season the inside with salt and pepper, and dip them in a little milk and then in Scotch oatmeal. Have ready some hot fat in a frying-pan and cook the fish for about 10 minutes till a nice golden brown. Drain on paper and serve hot.

Fish Cakes

4 to 6 oz. cold fish, 4 oz. cooked potatoes, 1 oz. dripping, 1 tablespoonful milk, salt and pepper, 1 egg, breadcrumbs. Average cost, ½d. each.

Remove the skin and bones from the fish, chop it finely, and mash the potatoes. Melt the dripping, mix in the potatoes, fish, and milk, add the seasoning, and form into flat cakes. Beat up the egg, brush the cakes over with it, cover with breadcrumbs, and fry a nice colour in hot fat. The cakes may also be coated with flour only, or with milk and oatmeal, as in the last recipe.

Steamed Fish Pudding

½ lb. cold fish, 4 potatoes, 1oz. dripping or butter, ½ gill of milk, 1 egg, salt and pepper. Average cost, from 6d. to 8d.

Wash, peel, boil, and mash four good-sized potatoes. Shred the fish finely, add the dripping or butter, and mix well together with the potatoes and milk. Beat up the egg, and stir in. Put the mixture into a greased basin, steam for 1 hour, turn out, and serve hot with white sauce or melted butter, or it may be put in a pie-dish and baked.

Baked Haddock

1 fresh haddock, 1 tablespoonful chopped suet, ½ teaspoonful herbs, 2 tablespoonfuls breadcrumbs, teaspoonful chopped parsley, 1 egg, salt and pepper. Browned breadcrumbs. Average cost, 8d. to 10d.

Wash, scale, and dry the fish; remove the eyes. Prepare a stuffing with the suet, breadcrumbs, herbs, parsley, and season with salt and pepper. Moisten the stuffing with half the egg. Stuff the body of the fish and sew up the opening. Pass a trussing needle, threaded with string, through the tail of the fish, centre of body, and the eyes, draw the fish up into the shape of the letter S, brush it over with the remainder of the beaten egg, and sprinkle over some browned breadcrumbs and a few pieces of dripping. Bake in a brisk oven from 30 to 40 minutes, basting frequently. Dish up, and serve with brown sauce, melted butter, or ordinary gravy.

Skate or Roker with Burnt Butter

1½ lb. fish, 1 bay leaf, 1 or 2 slices of onion, 1 dessertspoonful vinegar, 1 dessertspoonful salt, and 1½ oz. butter.

Procure the fish skinned, cut into strips and twist into rings. Skinning fish is rather a tedious operation; it is therefore advisable to get it skinned and tressed from the fishmonger. Put the fish into a stewpan or fish-kettle, with enough salted water to well cover it, add a bay leaf, the onion slices, and the vinegar. Bring it to the boil, remove the scum, and cook till tender. Take up the fish, drain well, and put it on a hot dish. Melt the butter in a frying-pan, allow it to get brown, then pour it quickly over the fish, sprinkle over a little chopped parsley, and serve.

Cod Steak with Bacon

Procure a slice of cod weighing from ½ to ¾ lb. Wash and dry it, and place on a buttered baking-tin. Prepare a mixture of breadcrumbs, savoury herbs, and parsley, and bind it with a little beaten egg. Place this upon the fish, and season with salt and pepper. Put a thin slice of lean bacon over the stuffing, and bake for about 20 minutes in a moderate oven. Dish up, and pour round it some anchovy sauce. Average cost, 7d.

Curried Fish Fritters

Remove the skin and bones from a smoked haddock, or kippered herring. Cut the fish into small pieces, and season with pepper and curry powder. Dip each piece into frying batter (see page 13, Fried Fish), and fry in hot fat to a nice brown. Drain well on paper, dish up, and serve. Average cost, from 4d. to 6d.

Herrings, Stuffed

4 fresh herrings, 2 tablespoonfuls breadcrumbs, 1 tablespoonful milk, 1½ tablespoonful chopped suet, 1 tablespoonful chopped parsley, ¼ teaspoonful powdered mixed herbs, salt, pepper. Average cost, 7d.

Wash, dry, and slit the herrings, remove the heads and backbones, and divide into halves. Season the insides well, spread on a little forcemeat made of the above ingredients, and roll up each fillet lightly, beginning at the head. Place in a greased pie-dish, add a few bits of dripping and a little stock or gravy, and bake from 25 to 30 minutes in a moderately hot oven.

Monks, Catfish, Gurnet and Bream

These are not very generally used, but can be relied upon to provide many a delightful dish, as a change to the better-known kinds, catfish and bream, in particular, being very succulent and nutritious.

Conger Eel

This useful fish has of late come into prominence, for it forms a food of no mean merit. In the great fish markets of Paris and Bordeaux, conger eel is much sought after. When boiled and served with good white sauce, or egg sauce, it forms a delicious dish. It can also be cut into slices, boned and skinned, and dipped in batter, or egg and bread crumbed, and fried in deep fat. Slices of conger eel an also be grilled and served with parsley butter—his makes an excellent dish.

Chapter 4

Meat Dishes

Roast or Baked Meat

Wipe and trim the meat, tie up if necessary, put in a baking tin (on a trivet if possible) with a little ripping and flour on top, and put in a very hot even for the first ten minutes, so as to harden the outside and to keep in the juices. The meat requires to be frequently basted. When the meat is browned, move it to a cooler part of the oven until it is quite one. The time allowed for baking is usually hour for each pound of meat and ¼ hour over; eat such as veal, pork or lamb, requires a little longer, 25 minutes to each lb. and 20 minutes over. When done, take up the meat and make the gravy. To do this, pour off nearly all the dripping in the tin, and keep in a jar for future use. Pour into the tin about a gill or more of hot water, season with salt and pepper, stir over the fire until it boils, and pour the gravy round the meat.

Yorkshire Pudding

Boiled Mutton

1 lb. scrag end of neck or breast of mutton, 2 small onions, 2 carrots, 2 turnips, 1 oz. pearl barley, salt. Average cost, 8d.

Clean (peel or scrape) the vegetables, cut the carrots and turnips into quarters, wash the barley, wipe the meat, and joint it if necessary. Have ready a saucepan with boiling water (about a quart), put in the vegetables, barley, and meat, add about a teaspoonful of salt, boil up, and skim. Let the whole simmer gently from to 1 hour. Remove the scum whilst cooking. Dish up, and put the vegetables round the meat, pour over some of the liquor and serve. The rest of the broth must not be thrown away, but used for soups and gravies.

Grilled or Broiled Meat

This is cooking meat on a gridiron over or in front of the fire. In broiling, as in roasting, the meat must be kept close to the fire for the first few minutes, to harden the outside and thus prevent the juices from escaping. Small pieces of meat, such as chops and steaks, are especially suited for this mode of cooking. The gridiron should be made hot, and then greased with a little fat or dripping before the meat is placed on it. Avoid using a fork for turning the meat, as this may cause the juices to escape.

Meat Roasted in the Pot

This way of roasting is especially suitable for small pieces of meat, and is far more economical because of the small quantity of fuel required. Melt one or two ounces of dripping in a pot (earthenware preferred). Brown the meat in this, so as to harden the outside and keep in the juices. Then draw the pot to the side of the fire and let the meat cook slowly with the lid on, basting it frequently. Time required, about 20 minutes to the pound, and 15 minutes over.

Fried Chop or Steak

Put the chop or steak in a frying-pan with sufficient dripping to prevent burning, and fry over a brisk fire. It will take from two to 12 minutes to cook; turn every 4 minutes. Dish up, pour off the fat, add a little stock or water to the pan, season with pepper and salt, stir until the gravy boils, and pour round the meat.

Note

A little Lemco or Oxo added to hot water makes an excellent gravy.

Irish Stew

1 lb. scrag of mutton, 2 lb. potatoes, 3 onions, about ½ pint water, pepper and salt. Average cost, 9d.

Wash and peel the potatoes, cut them into thick slices, peel and slice the onions, cut the meat into neat pieces. Place half the potatoes at the bottom of the saucepan, lay the meat and onions on them, sprinkle over some salt and pepper, add the water, and lay the rest of the potatoes on the top, keep the lid of the saucepan closely shut, and simmer gently for 1½ to 2 hours. Shake the saucepan occasionally to prevent the contents from burning.

Stewed Ox Kidney

1 lb. ox kidney, 3 oz. fat bacon, 2 small onions, ½ pint water, pepper and salt, ½ oz. flour (brown). Average cost, 10d.

Scald the bacon, after cutting it into small pieces. Wash and slice the kidney. Put it in a saucepan with the bacon and onions cut up small. Add the water, and simmer gently or about 2 hours. Mix the flour, pepper, and salt with 2 tablespoonfuls of cold water; add carefully to the stew, stir and cook for 5 minutes. Serve with sippets of fried bread.

Stewed Beef and Savoury Balls

½ lb. lean beef, 1 small onion, carrot and turnip, tablespoonful flour, 1 oz. dripping, pepper and salt, 1 pint water. Average cost, 5½d.

Melt the dripping in a saucepan, peel and slice the onion, brown it in the dripping, shake in the flour, and brown that also. Then add the water, meat, and vegetables, cut into pieces, and the pepper and salt. Simmer slowly for about 2 hours.

Savoury Balls

3 tablespoonfuls flour, ½ teaspoonful parsley, ½ teaspoonful herbs, 1 oz. suet, pepper and salt. Average cost, 1d.

Chop the suet, parsley and herbs, and mix them all together with the flour and seasoning. Add sufficient cold water to make a stiff paste. Divide it into eight pieces, roll each into a small ball, roll lightly in flour, and cook for ½ an hour in the stew.

Stewed Rabbit

1 rabbit, 4 slices bacon, 1 onion, 1 tablespoonful flour, 1 pint of water, salt and pepper to taste. Average cost, is.

Wash, wipe, and dry the rabbit after cutting it into pieces. Heat the saucepan and fry the bacon. Put it on a plate. Fry the onion. Mix together the flour, pepper and salt, and roll the pieces of rabbit in this mixture. Fry them in the bacon fat. When brown, remove and add the rest of the flour, brown it a little, season, and add the water. Boil up, stir well, and put in the meat. Simmer gently for about 1½ hours. Re-heat bacon, and serve. Savoury balls may also be served with this stew.

Steak Pudding

½ lb. flour, 4 oz. suet, ½ teaspoonful of baking powder, 1 gill cold water, 1 lb. pieces of steak, salt and pepper. Average cost, 11d.

Cut the meat into thin slices and dip into a little seasoned flour on a plate. Roll up, with a piece of fat inside. Chop the suet finely, mix with the flour a pinch of salt and the baking powder. Make into a stiff paste with cold water, roll out and line a pudding basin, keeping back 1/3 for the top, lay in the pieces of meat, and season. Pour in a little water, moisten the edges, and cover the top with paste. Tie up with a cloth, and boil or steam from 2 to 2 ½ hours.

Sea Pie

½ lb. lean beef or scraps, 1 small turnip, 1 carrot, onion, 2 potatoes, 5 oz. flour, 2 oz. suet, ½ teaspoonful baking powder, pepper and salt. Average cost, 7½d.

Wash and peel the potatoes, peel the onion and turnip, scrape the carrot, cut all into slices, and cut the meat into small pieces. Put the meat and vegetables in alternate layers in a saucepan, and allow to simmer gently with a little water. Mix the flour with the baking powder and a pinch of salt. Skin and chop the suet finely, add enough water to work into a stiff paste. Turn out on a floured board, roll out to the size of the saucepan lid, and put on the top of the stew. Cook slowly for about tis hours in all. When done, pass the knife round the edge of the saucepan to loosen the crust. Cut it in four pieces, take up and put on a plate. Put the meat and vegetables on a dish, lay the crust on top, and serve.

Baked Pig's Fry (Poor Man's Goose)

½ lb. pig's fry, ½ lb. potatoes, 1 onion, a few sage leaves, pint water, 1 teaspoonful chopped parsley, pepper and salt. Average cost, 4½d.

Cut up the fry, peel and chop the onion, put the meat into a pie-dish, sprinkle thickly with the onion, parsley, and sage, season with salt and pepper. Slice the potatoes and lay them on the top, put a few pieces of dripping over, fill up the dish with stock or water, and bake in a moderate oven for about 1 hour.

Note

Any cold meat may also be used up in this way, especially cold pork.

Stuffed Heart

1 sheep's heart, 1 small onion, 1 tablespoonful of soaked bread, 1 teaspoonful chopped sage, ½ teaspoonful sweet herbs, salt and pepper, dripping. Average cost, 6d.

Wash and trim the heart, make the sage and onion stuffing, and fill the heart with it. Cover with a greased paper, tie round with string, and bake in a tin in a hot oven for 1 hour. Baste frequently. The heart may also be roasted over the fire in a covered saucepan.

Stewed Tripe

1 lb. of tripe, 2 onions, pint of skim milk, ½ pint water, ½ dessertspoonful of flour, pepper and salt. Average cost, 8d.

Blanch the tripe, remove all fat, and cut into square pieces. Put the tripe, the chopped onions, and the milk, into a saucepan. Season with pepper and salt, and add ½ pint of boiling water or stock. Simmer gently for about 2 to 2½ hours. Blend the flour smoothly with a little cold milk, and pour in. Stir until it boils up, and cook for another to minutes. Dish up, and serve hot.

Liver and Bacon

½ lb. liver, 4 rashers of fat bacon, 1 tablespoonful of flour, pepper, salt, and a little fat. Average cost, 7d.

Wash the liver, dry it, and cut into rather thick slices. Mix the flour on a plate with one teaspoonful of salt, and half a teaspoonful of pepper, and dip each piece of liver in this. Cut each rasher of bacon into three or four pieces, fry till crisp in the frying-pan, take out and keep hot on a dish. Add a little fat, if necessary, to the bacon fat. When hot, put in the liver and fry quickly for about xo minutes, to a nice brown, then put the liver on a hot dish. Stir the remainder of the flour into the pan containing the fat, brown it nicely, add ¼ pint of boiling water, and pour it over the liver. Arrange the bacon round the dish. Serve hot.

Cornish Pasties

3 oz. meat, cooked or uncooked, 1 potato, 1 onion, 1 tablespoonful of water, pepper and salt. For the paste, ½ lb. flour, ½ teaspoonful baking powder, 3 oz. dripping, water to mix. Average cost, 1d. each.

Cut the meat, onion, and potato into small dice, and mix with the pepper, salt, and water on a plate. Rub the dripping into the flour, add a pinch of salt and the baking powder, and mix to a firm paste with a little cold water. Roll it out on a board, cut out some rounds, place a portion of the meat and vegetables on each, wet the edges and pinch together firmly, so that the join comes at the top. 'Work the edges between the fingers to form a kind of frill. Brush over with a little egg or milk. Place on a greased baking tin, and bake in a hot oven for ½ an hour. Cornish pasties proper are made with raw meat and vegetables.

Meat Patties

4 oz. beef or mutton, 6 oz. flour, 2 oz. dripping, ½ teaspoonful baking powder, salt and pepper. Average cost, ½d. each.

Cut meat into small dice, season with salt and pepper, and moisten with a little water. Put the flour, baking powder, and a pinch of salt into a basin. Rub in the dripping, add enough water to mix to a stiff paste, roll out, cut into rounds, put 6 or more aside for covers. Roll out the trimmings, and stamp out 6 or more linings for the patty-pans, which must be ready greased. Put a dessertspoonful of meat in each, wet edge of paste, put on the covers, press edges together, and make a hole on the top to let out steam. Decorate with little leaves, etc., made from the scraps of paste left over. Brush over with milk or beaten egg, and bake in a fairly hot oven for about ½ an hour.

Steak Pie

1 lb. beefsteak or skirt, ¼ lb. bacon, salt and pepper, ½ lb. flour, 3 or 4 oz. dripping, ½ teaspoonful of baking powder. Average cost, is. 4d.

Put the flour in a basin with a pinch of salt and the baking powder, rub in the dripping. Mix to a firm paste with cold water, or make up as flaky pastry by rubbing in 1 oz. fat, and rolling in the remainder. Place the pie-dish upside down on the paste, and cut it ½ an inch larger on all sides, moisten the rim of the dish with water, and put on some strips of paste. Arrange the meat in layers, with thin slices of bacon, or dip each slice in seasoned flour and roll up neatly with a piece of bacon inside, season with salt and pepper, and half fill the dish with water. Wet the edge again, lay on the cover, press down the edges lightly, and trim off the paste. Decorate the pie with leaves, etc., brush over with beaten egg, and bake for 1 ½ hours.

Rabbit Pie

Cut a small, skinned rabbit into neat joints, and lay the pieces in a pie-dish with alternate layers of thin pieces of bacon. Make ½ lb. of flaky pastry, or ordinary short crust, and follow the directions for Steak Pie. Average cost, is. 4d.

Meat Cakes

½ lb. cooked meat or sausage meat, ½ lb. cold potatoes, 1 oz. dripping, a tablespoonful of milk, pepper and salt. Average cost, 7d.

Mince the meat finely, mash the potatoes or rub through a sieve, moisten with the milk, season with pepper and salt,' and mix well together with a small piece of dripping, previously melted. Fill up one or two greased tin moulds, and bake for about 25 minutes; or form the mixture into small, round, flat cakes, dip them in flour, brush over with egg or milk, and coat them again with flour, or breadcrumbs if an egg *is* used. Then fry in hot fat until nicely browned. If sausage meat is used it will require a little longer.

Curry of Cold Meat

½ lb. cooked meat, 1 oz. dripping, 1 onion, ½ sour apple chopped finely, 1 tablespoonful of curry powder, 1 teaspoonful of flour, ½ pint of stock or water, ¼ teaspoonful of salt. Average cost, 7d.

Mince the onion, fry it brown with the apple in the hot dripping, add the curry powder and flour, stir, and cook for

2 minutes, and add the stock or water. Boil for a few minutes, put in the cold meat, cut into dice, add a little salt, simmer for ½ an hour, and serve with boiled rice. A good breakfast dish may be made by using hard-boiled eggs instead of meat for the curry. Allow one egg for each person. Curry proper should be made with raw meat.

Cottage Pie

1 lb. cold potatoes, ½ lb. cold meat, 1 small onion, 1 oz. dripping, 1 tablespoonful flour, 1 tablespoonful salt, ¼ table-spoonful pepper, a little water. Average cost, 8d.

Mash the potatoes finely with a fork, mix in a little seasoning and a tablespoonful of milk. Mince the meat, removing all fat, bone and gristle. Melt the dripping in a saucepan, chop the onion, and fry it in the hot fat. Add the flour, and brown nicely. Add about a gill of water and stir till it boils. Then season, and add the minced meat to the gravy, and heat through. Put the meat and gravy at the bottom of a pie-dish. Pile over the mashed potatoes, keeping them high in the centre to represent a thatched roof. Smooth over, and then score the surface with a fork. Sprinkle over some small pieces of dripping, and brown in a quick oven.

Bacon and Beans

Piece of fat bacon (2 or 3 lbs.), 1 peck of broad beans (measured in their shells), salt, water, 1 oz. butter, 1 oz. flour, ½ pint milk and water, 2 tablespoonfuls chopped parsley. Average cost, Is. 6d.

Soak the bacon for an hour in warm water. Cut away any discoloured bits, and take off the rind. Put the bacon into cold water, and bring it to the boil; when it boils, skim it and let it simmer gently, allowing ½ hour to each pound. Shell the beans, put them into boiling water in time to be finished when the bacon is ready, add 1 teaspoonful salt to every 2 quarts of water, and boil till tender. Young beans will take 15 minutes, old ones ½ hour. If dried haricot beans are substituted for fresh ones, cook until tender, between three or four hours, after soaking all night. Make ½ pint of melted butter sauce with the fat, flour, and milk and water, and add the chopped parsley. Serve the bacon on a hot dish, sprinkle over some browned crumbs, put the beans round the bacon, and pour the sauce over.

To Boil Rice for Curry

¼ lb. rice, 3 pints water, 1 teaspoonful of salt. Average cost, 1d.

Wash the rice thoroughly, throw it into a saucepan of boiling water with the salt, boil quickly for 10 to 15 minutes, skimming frequently. Strain off the water, and dry it in front of the fire until the grains are separated.

Chapter 5

VEGETABLES, ETC.

TO BOIL A CABBAGE

Take off all the faded leaves, cut the stalk across and put the cabbage to soak in a bowl of cold water and salt for an hour. Place it in a saucepan with enough boiling water to well cover it, add a good teaspoonful of salt, and as much carbonate of soda as will stand on a threepenny-bit. Boil without the lid, and very fast, for 20 to 30 minutes, till the stalk is soft. Drain it in a colander, and press with a plate to squeeze out the water. Cut it in quarters, and season with a little butter, salt and pepper before serving.

TO BOIL POTATOES

Scrub the potatoes, and peel them very thinly. Leave in cold water until wanted. Put them in a saucepan with enough warm water to cover them, and a teaspoonful of salt. Let them boil slowly for 20 minutes. Then try with a fork; if they are soft, pour away the water, stand the saucepan beside the fire, with the lid tilted, and shake occasionally to prevent the potatoes sticking to the pan and to make them dry and floury.

All root vegetables should be put into boiling water and boiled slowly, except old potatoes which should be put into cold water.

Baked Potatoes

Average cost, 2 lb. for 2d.

Wash and peel the potatoes, drain them well, and put them under the meat, in the same tin, having plenty of dripping to baste them. Sprinkle with salt, and serve them with the meat.

Fried Potatoes

Cut about 1 lb. of cold, cooked potatoes into slices; put them in a frying-pan containing 2 oz. of very hot dripping. Fry the potatoes a golden brown on both sides, being careful not to break them in turning. Sprinkle with pepper and salt before serving, and drain well on unprinted paper.

Stewed Potatoes

Boil 6 or 8 good-sized potatoes in their jackets; when cold, peel and cut them into slices. Put them in a stewpan with 2 oz. of butter or dripping, a pinch of savoury herbs, and 1 gill of milk. Season with salt and pepper, and add a little chopped parsley. Simmer gently for about 10 minutes, stirring from time to time to prevent burning. Average cost, 4d.

Potato Croquettes

4 large potatoes, 1 egg, 1 tablespoonful of flour, salt and pepper, breadcrumbs. Average cost, 3d. Boil the potatoes in their skins, then peel and mash them finely. Stir in the flour, and season with salt and pepper. Beat the mixture well. Form into croquettes about three inches long, brush with beaten egg, and roll in breadcrumbs. Fry in hot fat to a golden colour. Drain, dish up, and serve.

Boiled Onions and Sauce

Select large onions all the same size, peel them, cover with boiling water, boil for a few minutes, then drain them and return to the pan. Cover again with boiling water, add salt to taste, cook gently till quite tender, but not broken, for 2 or 3 hours. Drain well, and serve on a hot dish with white sauce or gravy poured over them. Average cost, 2d.

Mashed Turnips

Peel 1 lb. of turnips thickly. Cut them into quarters, boil till quite tender in salted water, drain and mash them. Add 2 tablespoonfuls of milk or a small piece of dripping, season with salt and pepper, and mix all thoroughly together. Dish up, and serve hot.

Colcannon

Take equal quantities of cold, boiled potatoes and cold, boiled cabbage. Mash the potatoes with a fork, chop the cabbage, mix both together, place them in a saucepan or frying-pan with a little dripping, season with pepper and salt, and stir over the fire till the vegetables are hot and slightly browned. Grease a basin or pie-dish, put in the mixture, and bake in a hot oven for about ½ an hour, or, if liked, the vegetables may be served immediately after frying.

Haricot Beans

½ pint haricot beans, 1 rasher of bacon, 1 small cooked onion, 1 oz. dripping, a little chopped parsley, pepper and salt. Average cost, 4d.

Soak the beans overnight in cold water with a small piece of soda. Drain off the water, put them in a saucepan with the dripping. When thoroughly hot, cover with cold water, cook slowly, and remove any scum that rises. Shred the onion. Cut the bacon into strips, fry in a frying-pan, put in the onion, and stir over the fire for a few minutes. Add this to the beans, season with pepper and a little salt, cook slowly till tender, stirring occasionally to prevent burning. Pile on a hot dish, and sprinkle with chopped parsley.

Stewed Mushrooms

½ lb. fresh mushrooms, 1 oz. butter, 1 dessertspoonful flour, 1 teaspoonful chopped parsley, seasoning. Average cost, 5d.

Remove the stalks, peel the mushrooms, fry them with the butter over a quick fire for a few minutes, season with pepper and salt, and put them on a plate to keep hot. Add the flour to the butter in the pan, fry a little, moisten with ½ pint of milk and water, boil, and stir for 5 minutes. Season to taste, put in the mushrooms, stew gently for 10 minutes. Dish up, sprinkle over with parsley, and serve hot.

LETTUCE SALAD

3 small lettuces, 1 handful of mustard and cress, a few slices of pickled beetroot, oil and vinegar, seasoning. Average cost, 4d.

Trim, wash, and drain the lettuces, wash the cress, break the lettuce leaves into small pieces, and drain all thoroughly by shaking them gently in a clean cloth. Season the salad with salt and pepper, and mix with oil and vinegar. Serve the salad in a fancy bowl, and garnish with slices of beetroot.

BEETROOT

Wash the beet well, taking care not to break or prick the skin, lest it should "bleed," or lose its colour whilst boiling. Put into boiling, salted water, and cook until tender. Remove peel quickly, slice up the beet, and send to table with white sauce.

STUFFED VEGETABLE MARROW

1 medium-sized marrow, ¼ lb. cold meat, ¼ lb. breadcrumbs, herbs, parsley, 1 egg, seasoning. Average cost, 9d.

Mince the meat and mix it with the breadcrumbs, seasoning, herbs, parsley, and bind together with the egg. Peel and cut the marrow in half, lengthways, and take out the seeds. Fill with the prepared mixture, and tie round with tape. Bake from ½ to ¾ hour, according to size, in a moderate oven.

Peas Pudding

1 pint split peas, 1 oz. dripping, salt and pepper. Average cost, 4d.

Soak the peas overnight, put them in a saucepan with a quart of cold water, boil about an hour, or until tender. Rub them through a sieve or colander, add the dripping, pepper and salt, and mix well. Flour a cloth, tie the pudding tightly in it, put it into boiling water, and boil for about 1 hour. A beaten egg may be added to the pudding, if liked.

Baked Tomatoes

6 tomatoes, 2 slices cooked bacon, 2 tablespoonfuls breadcrumbs, ½ oz. butter, ½ an onion, a little chopped parsley, salt and pepper. Average cost, from 6d. to 8d.

Boil and chop the onion very finely, chop the bacon finely, add the parsley, salt and pepper, with the remainder of the butter, and mix well with the crumbs. Scoop out a little of the inside of each tomato, and fill with the mixture. Put them into a greased pie-dish, cover with greased paper, and bake in a moderate oven for about 20 minutes.

Stewed Celery

1 head of celery, ½ pint milk and water, 1 oz. flour, salt and pepper, Average cost, 4d.

Trim and wash the celery, cut it into convenient lengths, put them with the milk and water in a saucepan. Cook gently for about 30 minutes. Thicken the milk with the flour, mixed smoothly with a little cold milk or water, add salt and pepper to taste, and continue to cook for another 10 minutes.

Potato Salad

Cut some cold potatoes into slices, and put them in a bowl. Add sufficient salad dressing, consisting of one part vinegar to two parts of oil, salt and pepper. Mix the whole carefully, pile up in the centre of the bowl, wipe sides, and sprinkle with chopped parsley, or finely chopped onion if liked. Lettuces or water-cress mixed with the above improve the salad. Average cost, 2 ½d.

Chapter 6

Breakfast and Supper Dishes

Porridge

2 oz. coarse oatmeal, 1 pint water, salt. Average cost, ½ d.

Put the water into a saucepan. When it boils, add the salt and sprinkle in the oatmeal, and stir till it boils and is smooth and free from lumps. Boil gently for 30 minutes. Serve with milk.

To Make Breakfast Coffee

2 tablespoonfuls of freshly ground coffee, ½ pint boiling water, ½ pint boiling milk, sugar to taste. Average cost 1 ½d.

See that the jug or coffee-pot is perfectly clean. Warm it, put the coffee in the jug, which should have a strainer, or, failing this, a muslin bag will do. Pour over the boiling water, cover well, set it on the top of the stove for 5 minutes. Serve with hot milk and sugar to taste.

Toad-in-the-Hole

Take 3 sausages or some pieces of cooked meat, chopped small, seasoned with pepper and salt. Average cost, 5d.

Boil the sausages for four minutes, cut them in halves lengthways, remove skin, and lay in a baking tin or greased pie-dish containing a little hot dripping. Pour over ½ pint of batter (see p. 59), and bake from to ½ of ¾ an hour.

Potato and Cheese-Cakes

3 large cooked potatoes, 1 oz. dripping or butter, 1 oz. grated cheese, ½ gill of milk, salt and pepper. Average cost, 3d.

Mash the potatoes smoothly, season with salt and pepper, add the butter or dripping (melted), half the cheese and the milk. Grease some little tin moulds, put a tablespoonful of the mixture in each. Smooth over with a knife, and sprinkle the remaining cheese on top. Bake in a fairly hot oven for about 15 minutes.

Potato Pie

1 oz. suet, 1 lb. potatoes, 2 onions, 2 oz. oatmeal, ¼ lb. flour, 1 oz. dripping, ½ teaspoonful baking powder. Average cost 4d.

Chop the suet finely, peel and slice the onions, wash and peel the potatoes and cut into thin slices. Fill a pie-dish in alternate layers of suet, onions, oatmeal, potatoes. Season with pepper and salt. Add ½ gill of water, and cover with a crust made of the flour, dripping, and baking powder. Bake for about 1 hour in a moderate oven.

Savoury Pudding

½ lb. pig's fry, 1 onion, 2 oz. suet, 1 teacupful of oatmeal, salt and pepper, ½ teaspoonful dried sage. Average cost, 5d.

Boil the fry for about 1 hour. Grate the liver and chop up the remainder of the fry, also the suet. Mix well with the oatmeal, pepper, salt, chopped onion and sage. Moisten with a ½ teacupful of the water in which the fry was boiled. Put into a greased basin, cover with a cloth, and boil, or bake in a greased tin, for about 1 hour.

Poached Egg

Half fill a frying-pan with water, add a pinch of salt, and ½ teaspoonful of vinegar. Average cost, 1½d.

Break a fresh egg into a cup, being careful not to break the yolk. As soon as the water boils slip the egg gently into the pan, and tilt the pan a little so as to keep the egg in shape. When set, take it up with a small slice, drain it and trim a little, then place it on a piece of toasted bread either dry or buttered.

Welsh Rare-Bit

2 oz. cheese, 1 oz. butter or dripping, pepper and salt, ½ teaspoonful made mustard, square of buttered toast. Average cost, 2d.

Grate or cut up the cheese small, put it in a saucepan with the butter or dripping, pepper, salt, and mustard. Stir for a few minutes till the mixture is thick and soft. Pour over the buttered toast, and serve very hot.

Macaroni Cheese

3 oz. macaroni, 2 oz. cheese, ½ oz. butter, ½ oz. flour, ½ pint skim milk, salt, pepper, and a little made mustard. Average cost, 6d.

Cook the macaroni in boiling water and a little salt for 4o minutes, then cut it into small pieces. Melt the butter in a saucepan, stir in the flour and add the milk, and boil for 5 minutes, stirring all the time. Season to taste, mix in the cooked macaroni and about two-thirds of the grated cheese, put it into a greased pie-dish, sprinkle the remainder of the cheese on top, and bake to a nice brown in a quick oven.

Savoury Rice

4 oz. rice, 2 oz. grated cheese, ½ pint of skim milk, 1½ oz. dripping, pepper and salt, breadcrumbs. Average cost, 4d.

Wash the rice, put it in a saucepan with boiling water slightly salted, boil up and drain; put it back into the pan, add the milk (hot) and about ½ pint hot water. Cook until tender. Stir in half the cheese, and enough seasoning to taste. Grease a pie-dish, or some little tin moulds, and line with breadcrumbs. Put in the cooked rice, sprinkle the top with grated cheese and breadcrumbs, and put a few tiny bits of dripping here and there, or the rice and cheese may be arranged in alternate layers. Bake in a hot oven for about 20 minutes, turn out, and serve.

BUTTERED EGGS

2 eggs, ½ oz. butter, 1 teaspoonful Oxo, 2 pieces of toasted bread. Average cost, 4d.

Beat up the eggs and the Oxo, and mix well. Melt the butter in a frying-pan, pour in the eggs and stir till it begins to set. Pile this on toast (buttered), and serve.

A dessertspoonful of milk added to the *egg* is an improvement.

CHEESE-CAKE MIXTURE

4 oz. castor sugar, 2 oz. butter, 1 lemon (grated rind and juice), 1 yolk of egg. Average cost, 4d.

Put all ingredients together into a saucepan, and stir until quite hot (not boiling) and thick. This mixture will keep for some time if kept in an airtight jar.

MINCEMEAT

½ lb. apples, 8 oz. raisins, 8 oz. sugar, 8 oz. currants, ¼ lb. figs, ½ lb. suet, ½ oz. ground ginger, ¼ lb. chopped candied peel, ½ oz. mixed spice, rind and juice of one lemon, ½ gill of brandy.

Peel, core, and chop the apples finely, and the lemon-rind; mince the figs; stone and chop the raisins, and clean the currants. Mix with the rest of the ingredients, and keep in a covered jar. Average cost, 8d. per lb.

Chapter 7

Sweets, Puddings, etc.

Arrowroot Pudding

1 tablespoonful of arrowroot, ½ pint of milk, 1 egg, 1 teaspoonful sugar, a little grated nutmeg, if liked. Average cost, 4d.

Put the arrowroot in a cup, mix with a tablespoonful of milk, put the remainder of the milk in a saucepan, and boil up. Stir the mixed arrowroot and the egg, well beaten, put the mixture in a greased pie-dish, and bake for about 15 minutes in a hot oven.

Note

For an invalid the yolk and white of the egg should be separated, and the latter whipped up stiffly before adding.

Fig Pudding

¼ lb. stale bread, ¼ lb. flour, 3 oz. suet, salt, ½ teaspoonful baking powder, 6 oz. figs, 3 oz. sugar, ½ gill milk, nutmeg. Average cost, from 4d. to 5d.

Soak the bread, chop the figs and the suet. Mix all the dry ingredients. Squeeze the bread and break it with a fork, add to the rest of the ingredients, and mix well. Put the mixture into a greased basin, cover with a cloth, and boil for 2 or 3 hours.

Tapioca or Rice Pudding

2 oz. tapioca or rice, 3 tablespoonfuls Nestle's condensed milk, ¾ pint water, 2 eggs. Cost, 4d.

Wash the tapioca and put it in a buttered pie-dish with a little milk, to soak for 10 minutes. Beat up the eggs and mix with the milk and water. Pour this over the tapioca, and bake slowly for about an hour.

Cambridge Pudding

¾ lb. flour, 1½ oz. castor sugar, 1 egg, ½ lb. apples, 1 pint water, 3 tablespoonfuls Nestle's condensed milk. Cost, 5d.

Dissolve the milk in the water, peel, core, and slice the apples, make a smooth batter of the flour, milk, and egg, add the sugar and the apples. Pour the mixture into a buttered pudding-basin, tie over a floured cloth, and boil for about 2 hours.

Semolina Pudding

2 oz. semolina, 1 pint skim milk, 1 oz. suet, 1 tablespoonful of sugar. Average cost, 3d.

Chop the suet finely. Put the semolina, suet, and sugar in a pie-dish and mix well, pour the milk over, and bake in a cool oven for

1½ hours. A little nutmeg may be grated over the top before baking, if liked. The suet should only be added when skim milk is used.

Golden Pudding

½ lb. flour, ¼ lb. treacle, 2 oz. suet or dripping, 1 gill milk, ¼ teaspoonful carbonate of soda, 1 teaspoonful ground ginger. Average cost, 4d.

Chop the suet and mix with the flour. Add the treacle and soda dissolved in the milk. Mix all together. Pour into a greased basin, cover with a greased paper, and steam for about 2 hours. An egg is an improvement.

Cabinet Pudding

¼ lb. stale bread free from crust, ½ pint milk, 1 egg, 2 oz. sugar, 2 oz. raisins or sultanas, butter to grease the mould. Average cost, 5d.

Stone the raisins (clean sultanas, if used), and put a few here and there on the side and bottom of a buttered basin or mould. Cut the bread into small squares, and with it fill the basin or mould lightly. Beat the egg, add sugar and milk, and pour over the bread. Let it stand long enough to soak the bread, cover with greased paper, put it in a saucepan with enough boiling water to come half way up the side of the basin. Steam for about 40 minutes, turn out, and serve.

Note

If a richer pudding is wanted, sponge cakes or biscuits may be substituted for the bread.

Macaroni Pudding

2 oz. macaroni, 1 pint of skim milk, 1 egg, 1 tablespoonful sugar, a little grated nutmeg or lemon-rind. Average cost, 4d.

Break the macaroni into short pieces, simmer it in the milk for ½ an hour, mix in the sugar, and, when a little cool, the egg well beaten, or the macaroni may be cooked separately and put straight into the pie-dish. Flavour with the nutmeg or lemon-rind, put into a greased pie-dish, and bake for about 15 minutes.

Lemon Pudding

½ lb. flour, 4 oz. chopped suet, 1 egg, rind and juice of 1 lemon, ½ teaspoonful baking powder, 3 oz. sugar, ½ gill milk. Average cost, 7½d.

Mix the chopped suet, flour, sugar, and grated lemon-rind in a basin. Beat up the egg, add the milk, and stir into the dry ingredients. Put the mixture into a greased basin, cover with a greased paper and steam 2 hours. The addition of bread-crumbs makes this pudding lighter.

Hasty Pudding

1 pint milk, ¼ lb. flour, salt, butter. Average cost, 3d.

Boil the milk and salt. Sprinkle in the flour, stirring briskly all the time. Boil until smooth and thick. Pour into a greased pie-dish. Place a piece of butter on the top, and serve at once.

Christmas Pudding

¼ lb. flour, ¼ lb. breadcrumbs, ½ lb. chopped suet, 4 oz. currants, ½ lb. sultanas or raisins (stoned), 6 oz. mixed peel (shredded), ¼ lb. castor sugar, 1 lemon, rind and juice, 3 eggs, milk to mix. Average cost, is 5d.

Mix all the dry ingredients. Stir in the beaten eggs, and sufficient milk to make the mixture rather moist. Boil in well-greased pudding-basin for about 4 hours.

Currant Fritters

3 eggs, 3 oz. flour, 4 oz. currants, 4 tablespoonfuls boiled rice, sugar to taste, a grate of nutmeg, ½ pint of skim milk, a pinch of salt, frying fat. Average cost, 7d.

Mix the yolks of eggs and the flour, and add the milk gradually till a smooth and light batter is obtained. Add the salt to the whites of eggs and whisk stiffly, stir them lightly into the batter, add the currants, rice, nutmeg, and enough castor sugar to sweeten. Drop the mixture in spoonfuls into hot fat, and fry a light brown colour. Drain the fritters on paper, dish up, and dredge with castor sugar.

Banana Fritters

Peel 4 to 6 bananas, cut each in half, longways, put in a pie-dish, and sprinkle over with lemon juice and castor sugar. Then dip in a light frying batter. Fry in hot fat to a light brown colour. Drain, dish up, and serve hot.

BLACK-CAP PUDDING

½ lb. flour, 2 eggs, ¾ pint of skim milk, 2 oz. currants, a pinch of salt. Average cost, 5d.

Sift the flour into a basin, add the salt, beat up the eggs and stir gradually into the flour, adding the milk by degrees, and work into a batter. Butter large or 2 small pudding-basins, and arrange the currants at the bottom. Pour over the prepared batter. Cover the basins with buttered paper, and steam for 1½ hours.

PLUM ROLY PUDDING

½ lb. flour, 3 oz. suet, a pinch of salt, 3 oz. of currants or sultanas. Average cost, 5d.

Chop the suet finely, mix with the flour and salt. Clean the currants and mix with the flour and suet. Make into a stiff paste with water or milk, roll out an oblong shape, then roll up, put into a floured cloth, and boil for 1½ hours.

BAKED ROLY PUDDING

½ lb. flour, 3 oz. dripping or butter, ½ teaspoonful baking powder, jam. Average cost, 4d.

Rub the dripping into the flour and mix with water to a stiff paste. Roll out on a floured board and spread with jam. Roll it up, and bake on a greased tin for ¾ of an hour in a moderate oven.

Fruit Pie

¼ lb. flour, 2 oz. dripping, ½ teaspoonful baking powder, 1 lb. apples or other fruit, 1 teaspoonful castor sugar, 3 oz. moist sugar.

Peel, core, and cut the apples into small pieces, place them in a small pie-dish, add the moist sugar, and pour in a little water. Mix the flour and baking powder, rub in the dripping lightly, add a pinch of salt and the castor sugar, and mix with cold water to a stiff paste. Roll it out, line the edge of the dish (see Steak Pie), moisten the edges, put on the cover, mark the edges with the back of a knife. Brush the top with a little water, dredge over with castor sugar, and bake in a fairly hot oven for about ½ an hour.

Bread-and-Butter Pudding

3 or 4 thin slices of bread and butter, ½ pint milk or 2 tablespoonfuls Nestle's condensed milk mixed with ½ pint water, 2 oz. cleaned currants, 1 dessertspoonful sugar, 1 egg. Average cost, 4½d.

Fill a buttered pie-dish with alternate layers of bread and butter and currants. Beat the egg in a small basin, add the sugar and milk, and pour over the bread, etc. Bake in a moderate oven for about ½ an hour. Jam or marmalade may be used instead of currants.

Cocoa Pudding

1 tablespoonful Brown & Polson's patent cornflour, 1 tablespoonful cocoa, 1 tablespoonful castor sugar, ¾ pint milk, 1 egg. Average cost, 5d.

Mix the cornflour and cocoa smoothly with a little milk. Boil the remainder in a white saucepan, pour it over the cornflour, return to the saucepan, and stir well until it boils. When cool, add the sugar, yolk of egg, and beat well. Flavour, if liked, with a few drops of vanilla. Beat the white of egg to a very stiff froth, stir lightly in, pour into a buttered pie-dish, and bake in a moderate oven for about 25 minutes.

Marmalade Pudding

4 oz. flour, 4 oz. breadcrumbs, 4 oz. suet, 4 oz. sugar, 1½ good tablespoonfuls marmalade, 1 egg, 1½ gills skim milk, ½ teaspoonful baking powder. Average cost, 7½d.

Chop the suet finely, mix all the dry ingredients. Add the marmalade, egg, and milk, stir well, pour into a greased pudding-basin, cover with greased paper, and steam for 2½ hours. Serve hot, with marmalade sauce round.

Marmalade Sauce

1 gill water, 1 dessertspoonful castor sugar, 1 dessertspoonful marmalade. Average cost, 1d.

Put the sugar and water into a clean saucepan. Boil for 10 minutes, take off the scum as it rises, add the marmalade, boil again, and pour round the pudding.

Baked Custard

2 eggs, ¾ pint best milk, 1 dessertspoonful sugar, a little butter, a grate of nutmeg. Average cost, 5d.

Break the eggs separately into a basin, add the sugar, beat well, add the milk by degrees. Pour into a buttered pie-dish, grate the nutmeg over, and bake slowly in oven till well set.

Pancakes

Prepare a batter as described on page 45. Melt a small piece of lard in a small frying-pan. When hot, just cover the bottom of the frying-pan with batter. When the under surface is pale brown, turn it over and brown the other side. Drain each pancake on paper, dredge with castor sugar, roll up and keep hot while the remainder of the pancakes are being fried. Serve with lemon. Cost, about 4d. for 8 pancakes.

Six-Cup Pudding

1 teacupful flour, 1 teacupful chopped suet, 1 teacupful breadcrumbs, 1 teacupful sugar, 1 teacupful raisins and currants, 1 teacupful of skim milk. A small pinch of carbonate of soda, 2 teaspoonfuls vinegar, egg. Average cost, 8d.

Mix the flour, suet, breadcrumbs, and sugar together. Stone and cut up the raisins. Clean the currants, and mix the fruit with the other ingredients. Slightly warm the milk, and dissolve the soda in it. Add the vinegar to it, and while it is frothy, mix with the dry ingredients. Put the mixture into a greased basin and boil not less than 2 hours.

Baked Apple Dumplings

6 oz. flour, 2 oz. dripping, ½ teaspoonful baking powder, 4 apples, moist sugar. Average cost, 5d.

Mix the flour, baking powder, and a pinch of salt, rub in the dripping, and add enough water to make a smooth paste. Peel and core the apples. Roll out the paste about a ¼ of an inch thick, cut out 4 rounds, each large enough to enclose an apple. Put 1 teaspoonful of moist sugar and 1 apple on each round of paste. Wet the edge of the paste, fold over until the apples are quite covered. Take care that the edges are well joined, brush over with water, sprinkle with castor sugar, place on a greased baking sheet, and bake for about 25 minutes.

Banana Batter Pudding

Prepare a light, smooth batter with 2 eggs, pint of milk, and 4 oz. of flour (sifted), and pour into a buttered basin. Peel and slice 6 bananas, and put these into the prepared batter, mix with 2 oz. of castor sugar, cover with a greased paper, and steam for about 2 hours. Turn out, and serve with custard.

Banana Custard Pudding

Half fill a pie-dish with peeled and sliced bananas, and dredge over with castor sugar. Make 1 pint of custard with 2 eggs, ¾ pint of milk, and 1 oz. sugar, and pour over the bananas. Put a few very small bits of butter and a grate of nutmeg on top, and bake in a fairly hot oven for about ½ an hour.

Banana Trifle

Cut 4 sponge-cakes into slices, and put a layer in the bottom of a pie-dish, then a layer of peeled and sliced bananas. Sprinkle with lemon-juice, then add another layer of sponge-cake and more slices of bananas, until the dish is filled. Pour 1 pint of hot custard over all, and serve cold. Cost, about 9d.

Banana Pudding (Hot)

Skin and slice 6 bananas. Mix 2 tablespoonfuls of Brown & Polson's cornflour with a little cold milk, and stir into 1 pint of boiling milk, add 2 tablespoonfuls of castor sugar and 2 eggs. When the mixture has thickened add the bananas, and pour all into a pie-dish (buttered), and bake in a fairly hot oven for about 20 minutes. Probable cost, 9d.

Baked Banana Pie

Cut some thin slices of bread, butter them, and put them in a buttered pie-dish. Peel 6 bananas, cut them in half, lengthways, and fill the dish alternately with slices of bread and banana. Beat up 2 eggs, add 2 tablespoonfuls of castor sugar, and add a grate of nutmeg, stir into this 1 pint of boiling milk, and pour all into a buttered pie-dish. Bake for about 25 minutes, and serve hot. Cost, about 8½d.

Baked Bananas

Peel 6 bananas, and cut each in half, lengthwise. Dissolve 1 tablespoonful of butter, and mix with 2 tablespoonfuls of castor sugar and the juice of a lemon. Range the bananas neatly in a buttered pie-dish, and coat each layer of fruit with the sweetened butter, then bake in a fairly hot oven for about 25 minutes. Baste the fruit occasionally, and serve hot or cold. Probable cost, 7d.

Stewed Bananas

Put 6 bananas on a greased baking sheet, in their skins, and bake them in a moderate oven for about an hour, then peel them. Make a syrup with 4 oz. of loaf-sugar and a pint of water, and flavour with a little lemon-rind. Pour this over the bananas, dish up, and serve hot or cold. Cost, about 7d.

Apple Charlotte

4 large apples, ¼ lb. breadcrumbs, 2 tablespoonfuls sugar, 1 oz. dripping. Average cost, 6d.

Peel the apples thinly, cut them in quarters, take out the cores, and cook in a little water until tender. Grease a pie-dish, line it with breadcrumbs, and fill it with layers of stewed apples and sugar, and a little grated nutmeg. Cover the top with breadcrumbs, put the dripping in little pieces on them, and a piece of greased paper on the top. Bake slowly for an hour. Turn it out of the dish to serve, and sprinkle the top with castor sugar.

Coffee Mould

1 pint of milk, 2 oz. patent cornflour, 1 oz. castor sugar, coffee essence. Average cost, 4d.

Blend the cornflour with a little cold milk, and stir it into the rest of the milk when boiling. Boil for 8 minutes, stirring well. Take off the fire, add the sugar and sufficient coffee essence to flavour. Pour into a mould rinsed with cold water, and leave to get cold.

Cornflour Blancmange

2 oz. of patent cornflour, 1 pint of milk, 1 oz. of castor sugar, 1 strip of lemon-rind. Average cost, 3d.

Blend the cornflour smoothly with a little cold milk, bring the remainder almost to the boil, pour in the blended cornflour, add the sugar and lemon-rind, and stir and boil until the mixture leaves the sides of the saucepan. Remove the lemon-rind, and pour into a mould that has been rinsed with cold water and left wet. Let this stand till cold and firm. Turn on to a dish, and place a spoonful of preserve at each end.

Banana Cream

3 bananas, ½ oz. gelatine, 2 eggs, 3 tablespoonfuls Nestle's condensed milk, ¾ pint water. Cost, 7d.

Mix the milk and water well together, add the gelatine, and let it soak for an hour, then stir over the fire until thoroughly dissolved. Add the beaten yolks of the eggs, and stir until the mixture thickens. When nearly cold, add the bananas, sliced thin, and the whites of the eggs, well beaten, and pour into a wetted mould.

Egg Jellies

2 eggs, 3 oz. loaf-sugar, ¾ oz. sheet gelatine, 2 lemons, the juice made up to 1 pint with water. Average cost, 7d.

Peel the lemon-rind as thinly as possible, simmer it gently in a little water for 20 minutes, strain and preserve the water. Squeeze the lemons, and make the juice up to 1 pint with the water, and in it dissolve the gelatine. Beat the eggs, add the sugar, lemon-juice, gelatine, and water, whisk over the fire, but do not boil. Strain through muslin. Set in little moulds, and turn out when cold.

Milk Jelly

1 pint of milk, ½ oz. of gelatine, 1½ oz. of castor sugar, a small piece of lemon-rind. Average cost, 4d.

Put the milk, gelatine, sugar, and lemon-rind in a saucepan. Heat gently until the gelatine is dissolved. Strain. Let this cool in a basin until commencing to set, then pour it in a mould, and stand in a cool place. When firm, turn out and serve. The milk must be stirred occasionally in the mould or else the cream will sink to the bottom when the mould is turned out.

Rice Mould

4 oz. rice, 3 tablespoonfuls Nestle's condensed milk, 1 oz. castor sugar, 1½ pints water, flavouring to taste. Cost, 3d.

Wash the rice and blanch it in water, then drain and place it in a saucepan with the water and milk, and simmer till thoroughly cooked. Allow to cool a little, then pour into a wetted mould. Turn out when set, and serve with jam or stewed fruit.

Batter for Toad - in - the - Hole, Yorkshire Pudding, and for Pancakes

1 egg, ¼ lb. flour, ½ pint of skim milk, salt. Average cost, 3d.

Put the flour and salt into a basin, add the egg and half the milk, and stir into a smooth paste. Beat it for ten minutes with a wooden spoon, then add the remainder of the milk.

Yorkshire Pudding

Pour ½ pint of the above batter into a Yorkshire pudding tin containing 1 oz. of melted dripping. Bake for about ½ an hour in a fairly hot oven.

Boiled Batter Pudding

Prepare ½ pint of batter, as above directed, and pour it into a greased basin, tie a cloth over, and boil 1 hour. The basin must be quite full.

This pudding may be made either sweet or savoury; in the former case, add a little sugar and serve with fruit, jam, honey, or golden syrup.

Chapter 8

Cakes and Pastries

Milk Scones

½ lb. flour, 1 level teaspoonful cream of tartar, ½ level teaspoonful carbonate of soda, 1½ oz. of butter, 1 teaspoonful of castor sugar, ¼ pint of milk. Average cost, 3d.

Sift the flour, cream of tartar, and carbonate of soda together. Rub the butter into the flour, add the sugar and mix to a paste with the milk. Roll out, and cut into fairly thick rounds with a small round cutter, or a wine-glass. Bake in a good oven for about 15 minutes. Brush over with milk or egg 5 minutes before taking out of the oven.

St George's Cake

4 oz. butter, 3 eggs, ½ lb. flour, 1 teaspoonful baking powder, 4 oz. sugar, ½ teaspoonful lemon-rind (grated), ¼ teacupful water, 1 tablespoonful Nestle's condensed milk. Cost, 10d.

Cream the butter and sugar. Mix in the egg and stir in the flour, baking powder, and lemon-rind lightly. Dissolve the milk in the water, and add this to the other ingredients last of all. Place the mixture in a lined tin, and bake in a moderate oven for about 1 hour.

Plain Cake

½ lb. flour, ½ teaspoonful baking powder, pinch of salt, 3 oz. dripping, 2 tablespoonfuls sugar, about 1 gill of milk. Average cost, 4d.

Mix the flour, salt, and baking powder well together, rub in the dripping, add the sugar and enough milk to mix stiffly. Pour into a greased cake-tin, and bake 45 minutes in a moderate oven.

Currant Cake

To make a currant cake add ¼ lb. cleaned currants or sultanas. The same mixture can be made into buns by adding less milk, and making a stiffer mixture.

Rice Cakes or Buns

¼ lb. ground rice, 3 oz. sugar, 2 oz. flour, 2 oz. butter, ½ teaspoonful baking powder, 2 eggs. Average cost, 6d.

Beat the butter to a cream, add the sugar, stir well, and work in the eggs one at a time; lastly, stir in the ground rice and flour and baking powder. Put the mixture into one large well-greased cake-tin, or several greased patty-pans. Bake for about ¾ of an hour if large cake, or 20 minutes if small ones. A few drops of flavouring or a little grated lemon-rind may be added to the mixture, if liked.

SEED BUNS

½ lb. flour, 2 oz. dripping, 1 teaspoonful carraway seeds, 2 oz. sugar, 2 tablespoonfuls of milk, 1 egg. Average cost, 3d.

Put the flour in a basin, rub in the dripping, add the sugar and carraway seeds, moisten with the milk and beaten egg, and mix well together. Place the mixture in little lumps on a greased baking tin, and bake for 15 minutes in a hot oven.

CURRANT BUNS

Make as seed buns, omitting the carraway seeds, and add instead 2 oz. of currants.

GINGERBREAD

½ lb. flour, 2 dessertspoonfuls treacle, 2 oz. dripping, 2 oz. brown sugar, 2 teaspoonfuls ginger, ¼ teaspoonful carbonate of soda, 1 tablespoonful of milk. Average cost, 4d.

Mix the flour with the ground ginger. Melt the fat, sugar, and treacle together in a saucepan, stir them into the flour and beat well. Last of all, add the soda dissolved in a little milk. Pour into a well-greased tin, and bake ½ an hour in a very moderate oven, as this mixture easily catches.

Oatmeal Biscuits

3 oz. flour, 3 oz. oatmeal, 2 oz. dripping or butter, ½ teaspoonful of baking powder, a pinch of salt, 2 oz. sugar, 1 tablespoonful of water. Average cost, 4d.

Mix the oatmeal and flour, rub in the dripping, add the sugar, salt, and baking powder, and mix to a stiff paste with the water. Flour the board, roll out the dough, cut into round biscuits, and bake for is minutes in a moderate oven.

Potato Scones

½ lb. cold potatoes, 2 tablespoonfuls flour, 1 oz. dripping, 1 dessertspoonful milk, and not quite a teaspoonful of salt. Average cost, 2 ½ d.

Mash the potatoes, or rub through a colander or sieve. Warm the dripping and milk together, add the potato, stir well, work in the flour, and roll out on a floured board ½ an inch in thickness. Cut into squares or rounds, and bake on a greased tin in a quick oven for about 10 minutes.

Victoria Sandwich

2 eggs, their weight in flour and sugar, 1 teaspoonful of baking powder, 1 tablespoonful milk, jam. Average cost, 7d.

Beat up the eggs and sugar well, lightly stir in the flour and milk. Put the mixture into a round, greased tin, and bake in a quick oven for 10 minutes. Spread one half with jam slightly warmed, and place the other half on the top. Sprinkle with a little castor sugar, and cut into triangular pieces.

Queen Cakes

6 oz. flour, 3 oz. butter, 3 oz. sugar, 1 oz. currants or chopped almonds, ½ teaspoonful baking powder, few drops of flavouring, pinch of salt, 2 eggs. Average cost, 1d. each.

Clean the currants, lightly grease some small cake-tins, and place a few of the currants in the bottom of each tin. Beat the butter and sugar to the consistency of cream. Beat and add the eggs, and stir the remaining ingredients in as lightly as possible. Fill the tins, bake in a moderate oven from 10 to 15 minutes, and cool on a sieve.

Shortbread

1 lb. flour, ½ lb. butter, ¼ lb. castor sugar, 1 teaspoonful ground rice. Average cost, 11d.

Mix all the ingredients together in a basin until a crisp paste. Roll out, cut into fancy shapes, and bake on greased paper in a very moderate oven for about 10 to 15 minutes.

Sally Lunn

1 lb. flour, 1½ oz. butter, ½ pint lukewarm milk, 1 egg, pinch of salt, ½ oz. German yeast. Average cost, 4½d.

Melt the butter in the milk, and cream in the yeast. Mix together. Add the salt to the flour, and stir in the yeast, milk, and the egg well beaten. Mix together till quite smooth and free from lumps. Put the dough in well-greased tins, and leave to rise for 1 hour. Then bake in hot oven for 15 minutes.

Hovis Scones

1 lb. Hovis flour, 2 oz. butter, 2 oz. sugar, 1 teaspoonful cream of tartar, ½ teaspoonful bicarbonate of soda, 1 egg, milk to mix.

Mix well together the flour, cream of tartar, and soda. Rub in the butter. Beat up the egg, and dissolve the sugar in it. Add the flour, with enough milk to make into a stiff dough. Shape the dough into round scones with the hands, or with a cutter. Brush over the tops with milk or egg, and bake on a greased tin, in a quick oven, for about 15 minutes.

Hovis Currant Bread

1½ lb. Hovis flour, 2 oz. butter, 2 oz. sugar, ½ lb. currants, ½ oz. yeast.

Cream a teaspoonful of sugar with the yeast until the latter becomes liquid. Melt the fat in ¾ pint of tepid water, and when the latter is only at blood heat, stir it to the yeast. Make a well in the centre of the flour, and pour in the yeast, fat, and water. Stir with the hands until everything is well mixed to a moist dough. Knead in the cleaned currants. Leave the dough to rise in a warm place until twice the size, then make up into loaves, and place in greased or floured tins. Leave again to prove for about 20 minutes, then bake in a good hot oven until nicely browned.

Sponge-Cakes

2 eggs, 2 oz. castor sugar, 2 oz. flour, ½ teaspoonful grated lemon-rind. Average cost, 3d.

Whisk the eggs and castor sugar together to a stiff froth, lightly stir in the flour and lemon-rind. Grease some sponge-cake tins, coat the inside lightly with a mixture of flour and sugar, and half fill with the mixture. Bake in a moderate oven for about 15 minutes.

Treacle Tart

¼ lb. flour, ½ teaspoonful baking powder, 2 oz. dripping, 2 tablespoonfuls treacle, 2 tablespoonfuls breadcrumbs. Average cost, 3d.

Mix the flour with a pinch of salt, rub the dripping lightly in, add water to make a stiff paste, and roll out on a floured board. Line a greased tin or plate with the paste, fill the centre with treacle and breadcrumbs mixed together, cover with strips of the paste, and bake for about 20 minutes.

Baking-Powder Bread

12 oz. flour, 1 teaspoonful salt, 2 teaspoonfuls baking powder, 1½ gills of milk or milk and water. Average cost, 3d.

Sift the flour, baking powder, and salt together. Stir in the milk, and mix quickly and smoothly to a stiff dough. Divide in four. Cut each piece into a large and a small one. Make each into a ball. Put a small piece on a larger one, to form a cottage loaf. Place on a greased tin, and bake in a quick oven for 15 minutes. Brush over with egg or milk a few minutes before taking from the oven.

Chapter 9

INVALID DISHES

BEEF TEA (QUICKLY MADE)

¼ lb. lean beef (steak), 1 gill of water.

Shred the meat very finely, taking care to remove all particles of fat and skin. Put it into a pan, pour cold water over it and let it stand covered for 10 minutes, or longer if possible. Bring it to the simmering point, and allow to simmer slowly for about 10 minutes, then strain, add a small pinch of salt, and serve. Average cost, 3d.

Note

Beef tea should never be allowed to boil. Be careful to remove all fat whilst it is hot with a piece of soft paper. Another good way is to put the meat and water in a covered jar and cook in the oven, or stand the jar in a saucepan of boiling water, keeping it thus for 1½ hours or longer.

Egg Flip (for Invalids)

1 egg, 1 teaspoonful of castor sugar, 1 teaspoonful of lemon-juice, ¼ pint of milk. Average cost, 2½ d.

Separate the yolk and white of the egg, beat the yolk in a cup with the sugar and lemon-juice, add the milk, mix well, and pour into a glass. Beat the white stiffly, and heap it lightly on top.

Home-Made Lemonade

1 lemon, 1 oz. sugar, 1 pint boiling water Average cost, 1½d.

Peel the lemon very thinly, and put the rind in a jug. Pare off all the white pith of the lemon, cut the lemon across into thin slices, remove all the pips, and put the slices in the jug. Add the boiling water and sugar, allow it to stand till cool, strain and serve.

Barley Water

1½ oz. pearl barley, 1½ pints water, ½ lemon, sugar. Average cost, 1d.

Wash the barley in cold water, and blanch it by putting it into cold water and letting it boil for I minute. Put it in a saucepan with 1½ pints cold water, add the thinly cut lemon-rind, boil slowly for about 1 hour. Let it stand till cold, add the lemon-juice, and sugar to taste, then strain and serve. A little milk may be added, if liked. If used for babies, the lemon-rind and juice must be omitted.

ARROWROOT CUP

1 tablespoonful arrowroot, ½ pint water, 2 tablespoonfuls Nestle's condensed milk, flavouring, and sugar to taste. Cost, 1½d. Mix the arrowroot smoothly with a little water, boil the remainder with the condensed milk, arid pour over the mixed arrowroot. Return this to the saucepan, and cook over the fire for 10 minutes, stirring all the while to prevent it from burning. When cooked, add the desired flavouring, sugar to taste, and serve.

BARLEY OR GROATS GRUEL

Average cost, 1d.

Take a dessertspoonful of ground barley or groats, add to it a pinch of salt, and mix well with a little cold water into a very smooth paste. Add to this a ½ pint of boiling water or milk, pour into an enamelled saucepan, and simmer for 10 minutes. Keep stirring all the time with a wooden spoon. Sweeten to taste with sugar.

SUET AND MILK

Average cost, 2½d.

Skin and chop ½ oz. suet finely. Rinse a sauce-pan with cold water, put in the suet and r pint of new milk. Bring slowly to the boil, simmer for about 30 minutes, then strain, and serve hot in a tumbler.

BLACK-CURRANT TEA

Average cost, ½ d.

Put 2 tablespoonfuls black-currant jam and 1 pint of water into a clean saucepan, bring to the boil, then strain, when it is ready for use.

BEEF-TEA CUSTARD

Average cost, 4d.

Thoroughly beat up an egg, stir in r gill beef-tea, add a pinch of salt. Pour into a greased cup, cover with buttered paper, steam gently from 30 to 40 minutes. When done, turn on a plate or serve in the cup. Instead of using fresh meat, beef-tea may be made with a gill of boiling water, to which a teaspoonful of Lemco or Oxo has been added.

LINSEED TEA

1 oz. linseed, ½ oz. liquorice, 1 quart water, 1 oz. sugar-candy. Average cost, 1½d.

Wash the linseed. Put it into a pan with the water, liquorice, and sugar-candy. Simmer gently from ½ to 2 hours. Strain and use.

CUSTARD PUDDING (FOR INVALIDS)

To 1 oz. of cornflour add ½ an oz. of castor sugar, ½ oz. of butter, a pinch of salt, and stir in gradually a pint of milk. Average cost, 4d.

Mix thoroughly, and stir it over the fire till it boils, then add the yolks of 2 eggs, pour into a buttered pie-dish, and bake in a slow oven until lightly browned.

Junket

1 pint new milk, 1 teaspoonful prepared rennet, sugar to taste, nutmeg. Average cost, 2½d.

Make the milk lukewarm, and pour it into a deep glass dish. Stir in the rennet, and leave in a warm place until set. Grate over some nutmeg before serving.

Chapter 10

Sauces

Melted Butter Sauce

1 oz. butter, 1 oz. flour, ½ pint cold water, or milk and water, salt. Average cost, 1d.

Melt the butter in a saucepan, add the flour, stir and cook 3 minutes without browning, add the water, stir, and boil up. Add a pinch of salt and boil for about 10 minutes, stirring occasionally.

Onion Sauce

Make ¾ pint of plain melted butter sauce. Peel ½ lb. of onions, boil them in water till tender, drain, and chop them finely. Add them to the melted butter sauce, season with pepper and salt, and boil for a few minutes longer. Average cost, 1½d.

Anchovy Sauce

Average cost, 1½d.

For anchovy sauce, add 2 teaspoonfuls of essence of anchovy to ½ pint melted butter sauce.

Parsley Sauce

Average cost, 1d.

For parsley sauce, add 1½ teaspoonfuls of chopped parsley to ½ pint melted butter sauce.

Brown Fish Sauce

½ oz. butter, 1 tablespoonful flour, ¾ pint seasoned fish stock, 1 onion (peeled and sliced), 1 dessertspoonful of Bisto.

Melt the butter in a saucepan, and fry in it lightly the onion. Add the flour and the Bisto, and stir over the fire for a few minutes, then pour in the fish stock. Season to taste with salt and pepper, and boil for about 5 minutes, stirring continually. Strain and use as required.

Curry Sauce

1 oz. butter, 1 dessertspoonful curry powder, 1 dessertspoonful Bisto, 1 small onion (peeled and sliced), apple (chopped), ½ pint stock, lemon-juice.

Fry the onion in the butter, add the apple, curry powder, and lastly the Bisto. Stir over the fire for a few minutes, then add stock, a few drops of lemon-juice, and boil for about 10 minutes, stirring continually. This makes an excellent sauce for meat, chicken, or rabbit curry.

Bread Sauce

¼ pint breadcrumbs, 4 cloves, ½ oz. butter, ¾ pint milk and water, 1 small onion, 6 peppercorns, seasoning. Average cost, 2½d.

Peel the onion and stick the cloves into it. Put it into a saucepan with the peppercorns and milk, and leave by the fire to infuse for some minutes. Strain the milk, and add the crumbs. Simmer very gently for a few minutes, add the salt, butter, and a little pepper, and serve.

Brown Sauce (For Stews, etc.)

1 oz. butter or dripping, ½ oz. flour, ½ to ¾ pint stock or water, seasoning of pepper, 1 heaped-up teaspoonful Bisto, 1 small onion (peeled and chopped).

Melt the butter and fry in it the onion to a pale brown, then stir in the flour and Bisto. Add the stock or water, and allow to boil for about five minutes, stirring all the time, season with a pinch of pepper, strain, and use as required.

To Brown Flour (For Sauces, etc.)

Put ½ lb. of flour into a baking sheet lined with paper, put it into a warm oven and brown it thoroughly. It will require turning about with a spoon, and should be sifted after browning. This is very useful for stews or gravies, and gives them a good colour. Keep it in a tin.

Clarified Fat

Take any scraps and trimmings of fat, cut them into small pieces, and put them into a saucepan with enough cold water to cover. Leave to simmer for several hours, skimming occasionally, till all the water has evaporated and the pieces of fat are as clear as oil. Then strain off the pieces, and put the fat aside until wanted.

Salad Dressing

2 tablespoonfuls salad oil, 1 teaspoonful made mustard, 4 tablespoonfuls vinegar, 2 tablespoonfuls milk, 1 teaspoonful sugar, pinch of salt. Average cost, 2d.

Mix oil, drop by drop, with mustard, add vinegar gradually, then milk in the same manner, taking care that the sauce does not curdle.

Mint Sauce

2 teaspoonfuls finely chopped mint, 1 tablespoonful sugar, 1 gill of vinegar. Average cost, 1½d.

Wash the mint, and chop it very finely. Put it in a basin with the sugar, and leave to dissolve for a while. Pour over the vinegar, and it is ready for use.

Apple

2 or 3 large apples, ½ oz. bu
3d.

Peel, core, and cut up the ap
with enough water to prevent the
pulp. Then beat them up, or mas
taste, and the butter, and serve in

APPLE SAUCE

2 or 3 large apples, ½ oz. butter, sugar to taste. Average cost, 3d.

Peel, core, and cut up the apples, and put them in a saucepan with enough water to prevent them burning. Let them simmer to a pulp. Then beat them up, or mash through a sieve. Add sugar to taste, and the butter, and serve in a sauce-boat.

Chapter 11

MANAGEMENT OF GAS STOVES

Keep the stove and all utensils scrupulously clean from top to bottom, inside and out.

All loose fittings in the oven should be frequently removed and thoroughly cleansed. Strong hot soda and water, with scouring soap, should be used.

Pots and pans for boiling should be as broad and flat as possible, to utilise the heat to the best advantage. Keep them free from loose dirt outside.

Do not use a pan that has been on a coal fire without first scraping and scouring it.

Do not boil a quart of water when you only need a pint. It wastes gas. Do not let your pans or kettles boil over, or the burners will get choked and rusty, but turn down the gas as soon as the contents boil. This saves gas as well as saving the mess.

Heating the Oven

Let your oven get thoroughly hot before commencing to bake, and then put as much into the oven as possible, so as to save gas

After the first ¼ of an hour or 20 minutes, turn down the oven burners ; you will only spoil the food and waste the gas if you forget this. An exception must be made when cooking puff pastry. Keep up the full heat then all the time until finished.

The hottest part of the oven is that near the top, or under the solid shelf if that be in use. The best results will be obtained by putting the food on a grid shelf instead of on the solid one, which may cause the food to burn. *When opening the oven door* keep your head on one side to avoid the heat.

Meat or fowls should be hung or placed on a grid shelf. Always keep the drip - tin under the oven No water is necessary in this tin.

When stewing, or baking milk puddings, keep the gas turned down to avoid the contents boiling, and do not try to hurry, or you will spoil the food. Place them above the solid shelf.

Pastry can best be cooked at the top of the oven, or under the solid shelf.

Always turn down or turn out the gas the moment it is not required. Careful attention to this point will save much money in the course of a year.

Be careful not to turn on the oven tap in mistake for a boiling burner. Accidents have happened through the oven filling with gas and an explosion occurring when the oven was opened and lighted.

Always have your light ready and the oven door open before turning on the gas.

Used with reasonable care, gas for cooking is very economical, and it is always clean, convenient, and reliable.

Chapter 12

INEXPENSIVE FAMILY MEALS

COSTING FROM 3D. TO 7D. PER HEAD

Dinner No. 1. For 6 persons. Total cost, 1s. 6d.

Potato Soup (see p. 10)	4d.
Savoury Pudding (see p. 41—double quantity)	8d.
Apple Pancakes	6d.
	1s. 6d.

Apple Pancakes

Peel, quarter and core 3 or 4 small sour cooking apples, then cut them into very thin and small slices. Put the apples into a stew-pan with ½ oz. of butter, a small piece of whole cinnamon, 1 oz. of sugar, and 2 or 3 tablespoonfuls of water. Cover the pan and cook the apples just till tender. Care must be taken so as not to cook the apples to a pulp.

Now prepare a pancake batter as follows: Put 6 oz. of sifted flour into a basin, also oz. of castor sugar and a pinch of salt. Beat up 1 or 2 eggs, and mix these with ¾ pint of milk, and stir this gradually into the flour, etc. Beat all thoroughly with a wooden spoon till a smooth and light batter is obtained. Remove the cinnamon from the apples, and mix the latter, but without any of the liquor, with the prepared batter.

Heat a small omelet-pan over the fire, wipe it thoroughly with a cloth, and put in a little clarified butter or lard, just enough to well grease the pan ; then put in enough apple batter to cover the bottom of the pan, and fry' the contents, first on one side, then turn over the pancake and fry the other side. They should be a nice golden brown. Continue thus to cook the remainder of the batter, which however must be stirred up each time so as to mix the apples. Dredge the pancakes with plain or cinnamon flavoured sugar, dish them up, rolled or flat, and serve hot.

Dinner No. 2. For 6 persons. Total cost, 1s. 9d.

Poor Man's Goose (see p. 24—double quantity)	10d.
Mashed potatoes	3d.
Baked Oatmeal Pudding	8d.
	1s. 9d.

Baked Oatmeal Pudding

Mix in a basin 3 oz. of fine Scotch oatmeal with 2 oz. of wheaten flour, add a pinch of salt, and stir in gradually about a gill of milk, and work into a smooth batter. Put into a saucepan a pint of milk, 2 oz. of loaf or moist sugar, and a strip of lemon-rind; bring it to the boil, then stir in the prepared batter. Continue to stir until it boils, then simmer gently for about 8 minutes. Remove the lemon-rind and let the mixture cool a little, then stir in 1 or 2 tablespoonfuls of cream. Pour half the mixture into a buttered pie-dish, next put in a layer of stewed apples mashed up finely, then pour in the remainder of the mixture. Bake in a moderately heated oven for about 45 minutes. Serve hot with either honey, jam, or golden syrup.

Dinner No. 3. For 6 persons. Total cost, 2s.

Stewed Beef, etc. (see p. 22—double quantity)	1s. 2d.
Stewed Celery (see p. 37)	5d.
Semolina Pudding (see p. 46)	5d.
	2s. 0d.

Dinner No. 4. For 6 persons. Total cost, 2s.

Winter Soup (see p. 12)	3d
Fried Herrings (see p. 14)	1s. 3d
Boiled Rice and Currants	6d
	2s. 0d.

Boiled Rice and Currants

1½ pints milk, 1½ pints water, 6 oz. rice, 4 oz. sugar, 4 oz. currants.

Put all in a double saucepan, if procurable, and cook until the rice has become soft. If put in a single pan it will be necessary to stir frequently, and to add the sugar just before serving, as this lessens the risk of its burning. The currants must be well cleaned before using.

Dinner No. 5. For 6 persons. Total cost, 2s. 8d.

Boiled Mutton (see p. 20—double quantity)	1s. 3d.
Haricot Beans (see p. 34—double quantity)	8d.
Banana Batter Pudding (see p. 54)	9d.
	2s. 8d.

Dinner No. 6. For 6 persons. Total cost, 2s.

Scotch Barley Broth	10d.
Macaroni Cheese (see p. 42).	6d.
Sponge Cake Mould	8d.
	2s. 0d.

Scotch Barley Broth

1 lb. beef (without bone), ½ lb. carrots, ½ lb. turnips, ¾ lb. onions, 4 oz. barley, celery or celery leaves, parsley, pepper and salt, 2 qrts. water.

Cut the meat into small pieces, and put it in a saucepan with the barley, a little salt, and the water; bring it to the boil, and skim. Wash, peel or trim the vegetables, and cut them up small, then put them into the pan. Cook gently until all the ingredients are tender. Season to taste with salt and pepper; chop a little parsley finely, and add it last of all.

Note

Mutton can be used instead of beef, but the large proportion of bone and fat make it rather more expensive.

Sponge Cake Mould

Cut 3 to 4 sponge cakes into slices, and with these line the sides and bottom of a plain charlotte pudding mould. Put 1 tablespoonful of cornflour and 1 tablespoonful of arrowroot into a basin, and mix with sufficient cold milk to form a batter. Boil up ¾ pint of milk, add 1 oz. of castor sugar, and mix with the batter; put this into a stewpan, stir over the fire, and cook gently for 10 minutes. Add the yolk of an egg, cook a little longer without boiling, and lastly add 1 oz of currants, and if liked, a little vanilla essence. Let the mixture cool a little, and pour into the prepared mould. Put the mould in a cool place, turn out on to a dish when quite set, and serve plain or with jam sauce.

Dinner No. 7. For 6 persons. Total cost, 2s. 8d.

Lentil or Pea Soup (see p. 9—double quantity)	8d.
Meat Cakes (see p. 28)	8d.
Stewed Potatoes (see p. 32—double quantity)	4d.
Fig Pudding (see p. 45—double quantity)	1s. 0d
	2s. 8d.

Dinner No. 8. For six persons. Total cost, 2s. 3d.

Bone Soup (see p. 7).	6d.
Cheese and Potato Pie	1s. 0d.
Banana Cream and Jam (see p. 57-58)	9d.
	2s. 3d.

Cheese and Potato Pie

½ lb. cheese, 2 lb. potatoes, ¼ lb. breadcrumbs, 4 oz. rice, 4 oz. dripping or margarine, 2 eggs, salt and pepper.

Cook the rice and potatoes separately in slightly salted water. Grate the cheese or chop it finely. Peel the potatoes, boil and mash them, and mix 1/3 of them with the cooked rice, the cheese, the breadcrumbs, and 3 oz. of the dripping or margarine, eggs and seasoning. If not moist enough, add a little stock or gravy. Put all in a greased pie-dish, and cover with the rest of the potatoes. Put the remainder of the fat in small pieces over the top of the potatoes. Bake to a golden brown in a fairly quick oven.

Dinner No. 9. For six persons. Total cost, 2s.

Irish Stew (see p. 21—double quantity)	1s. 6d.
Fruit Pie (see p. 51—double quantity)	6d.
	2s. 0d.

Dinner No. 10. For 6 persons. Total cost, 2s.

Macaroni Soup (see p. 10—double quantity)	8d.
Cornish Pasties (see p. 26)	7d.
Ginger Pudding	9d.
	2s. 0d.

Ginger Pudding

½ lb. flour, ½ oz. ground ginger, ½ oz. baking powder, ¼ teaspoonful salt, 6 oz. chopped suet, ¼ lb. bread-crumbs, ½ lb. golden syrup, ¼ pint milk.

Mix the flour, ginger, baking powder and salt, rub in the suet and add the breadcrumbs. Melt the syrup by slightly warming it, mix it with the milk, and add these to the dry ingredients. Mix all thoroughly, and place in a greased tin or basin. Cover with a plate or greased paper, and steam or boil for about 2 hours. Turn the pudding out and serve hot. A little white sugar sprinkled over improves the appearance.

Dinner No. 11. For 6 persons. Total cost, 2s. 4d.

Fried Fish (see p. 13-14)	9d.
Steak Pudding (see p. 23— 1 ½ quantity)	1s. 0d.
Apple Charlotte (see p. 56).	7d.
	2s. 4d.

Dinner No. 12. For 6 persons. Total cost, 2s. 6d.

Yorkshire Cutlets	10d.
Cheese and Lentil Pie	8d.
Lemon Pudding (seep. 48—double quantity)	1s. 0d.
	2s. 6d.

Yorkshire Cutlets

Cut some cold meat, either beef, veal, mutton or pork, into slices, not too thinly. Trim each slice into a heart or cutlet shape, then put the meat on a dish, season with salt and pepper, sprinkle over a little finely chopped parsley, also a little vinegar, a tablespoonful of olive oil, and a few drops of Worcestershire sauce. Have ready some mashed potatoes, coat each slice of meat with a layer of potato purée, and shape neatly. Egg and crumb each cutlet carefully, and fry to a golden colour in shallow or deep fat. When done, take up, drain carefully and dish up neatly. Garnish with a few sprigs of parsley, and serve hot.

Cheese and Lentil Pie

½ lb. cheese, 6 oz. lentils, 4 oz. onions, 3 oz. bread crumbs, 1 ½ oz. dripping or margarine, parsley, salt and pepper, little stock or gravy.

Wash and pick over the lentils; peel and chop the onion, and cook in a little water with the lentils, stirring frequently, unless a double pan is procurable. Grate the cheese. Put in a basin with the partly cooked lentils and onions; add the breadcrumbs, pepper and salt to taste, and about 1 tablespoonful finely chopped parsley. Moisten with stock or gravy, and mix all to a soft pulp. Put the mixture in a greased tin or pie-dish; place the fat in small pieces over the top, and bake in a hot oven until firmly set, and of a nice brown colour. Serve hot plain or with gravy.

Dinner No. 13. Total cost, 2s. 6d.

Liver and Bacon (see p. 25 —double quantity)	1s. 1d.
Tomatoes with Macaroni	9d.
Cottage Pudding	8d.
	2s. 6d.

Tomatoes with Macaroni

Break up into small pieces about 4 oz. of macaroni, and cook in fast boiling milk and water (about ½ pint of each) containing a little salt and ½ a bayleaf. When done, drain, remove the bayleaf, and reheat the macaroni in a little butter and a little of the liquor, season with salt and pepper, add a heaped-up tablespoonful of finely grated cheese, and keep hot.

Next cut into fairly thick slices 4 good-sized ripe tomatoes, and season with salt and pepper. Beat up an egg, mix some breadcrumbs with a little grated cheese. Brush each slice of tomato with egg, then roll in bread-crumbs and cheese. Fry these in a shallow tin with just sufficient butter or dripping to prevent the tomatoes from burning.

Make a border with the cooked macaroni on a hot dish, and place neatly in the centre the cooked slices of tomato. Sprinkle over a little chopped parsley, and serve hot.

Cottage Pudding

¾ lb. stale bread (cut up small), 4 oz. flour, 4 oz. sugar, ½ teaspoonful each of cinnamon and mixed spice, 3 oz. suet (finely chopped), 2 oz. currants, and sultanas or raisins and peel, ½ saltspoonful of salt, 1 piled-up teaspoonful of Bird's Egg Substitute.

Soak the bread in water for 3 hours, or overnight. Squeeze dry, and with a fork beat out all lumps. Mix it with the other ingredients, and a gill of water. Put all in well-greased basin, cover, and steam from 2 to 3 hours. Turn out on a dish. Pour over ½ pint of Custard and serve hot.

Dinner No. 14. For 6 persons. Total cost, 2s. 7d.

Sea Pie (see p. 23—double quantity).	1s. 4d.
Steamed Parsnips	3d.
Hasty Pudding (see p. 48—double quantity)	6d.
Macaroni Cheese (see p. 42)	6d.
	2s. 7d.

Steamed Parsnips

Wash and peel the parsnips rather thickly. Cut them in quarters, or if large into eighths. Steam the vegetables (seasoned with salt) in a covered vessel until quite tender. Dish up; pour over, if liked, a little rich gravy, hot bacon fat or dripping. Serve hot.

Dinner No. 15. For 6 persons- Total cost, 2s. 6d.

Steak Pie (see p. 27).	1s. 2d.
Colcannon (see p. 34).	4d.
Tapioca Pudding (see p. 46—double quantity)	7d.
Stewed Fruit	5d.
	2s. 6d.

Stewed Fruit

2 lb. apples, or any other fruit in season, ¼ lb. sugar, flavouring.

Put about ½ pint of water and the sugar on to boil in a stewing-pan, and skim. Prepare and clean the fruit, and add to the boiling syrup. Simmer gently until the fruit is tender. Lemon-rind improves rhubarb and apples. 2 cloves may also be put with stewed apples.

Note

Dried fruit, such as figs, apples, prunes, apricots, etc., should be first well washed, then soaked in water over night, using the water for the syrup. The amount of sugar and water required will vary according to the fruit used.

Dinner No. 16. For 6 persons. Total cost, 2s. 10d.

Baked, Stuffed Haddock (see p. 15—double quantity)	1s. 8d.
Mashed Turnips (see p. 33)	3d.
Haricot Beans (see p. 34—1½ quantity)	about 4 ½ d.
Baked Apple Dumplings (see p. 54—1½ quantity)	6½ d.
	2s. 10d.

Dinner No. 17. For 6 persons. Total cost, 2s. 9d.

Dutch Pot Braise	2s. 2d.
New England Pudding	7d.
	2s. 9d.

Dutch Pot Braise

3 lb. neck of mutton, 1 oz. dripping, ½ large cabbage or 2 small ones, 8 large potatoes, 2 onions, peeled and sliced, salt and pepper.

Melt the dripping in a casserole or roasting pot. Wipe the meat, joint it, and trim it neatly, and put it in the pot, cover down closely, and let it cook gently in the oven for 1 hours. Meanwhile trim and wash the cabbage, and slice it into thin pieces; peel the potatoes and cut them into thick slices, and place these, with the cabbage, round the meat. Season to taste with salt and pepper, cover the pan, and cook gently for another hour or so. This stew must be cooked very slowly or it will burn; but it is a very appetising method if properly treated.

New England Pudding

1 lb. flour, 6 oz. finely chopped suet, 1 teaspoonful baking powder, ½ tin treacle.

Prepare a suet crust by mixing the flour, the chopped suet, baking powder and a pinch of salt, with sufficient cold water to make it into a fairly stiff paste. Divide the suet crust into two parts; roll out one part about s an inch thick, and line a greased pudding-basin with the paste, press it well in to make it fit smoothly. Roll out the remainder of the paste to the same thickness, and keep back a piece large enough to cover the top of the pudding-basin. Pour a tablespoonful of treacle (warmed) into the lined basin; cover it with a piece of paste cut to fit, and continue adding alternate layers of treacle and paste until the basin is nearly full. After the last layer of treacle, wet the edges of the pastry, cover with the lid that was set aside, press the edges together, cover with greased paper on a cloth, and steam or boil for 2 ½ hours or longer.

Dinner No. 18. For 6 persons. Total cost, 3s.

Fish Soup (see p. 8—double quantity)	1s. 2d.
Rolled Bacon Pudding	9d.
Potato Croquettes (see p. 33)	6d.
Baked Bananas (see p. 56).	7d.
	3s. 0d.

Rolled Bacon Pudding

Chop finely ½ lb. of suet, and mix it with 1 lb. of flour, add a pinch of salt, and enough water to make a smooth paste. Roll it out on a floured board, divide into 2 pieces, roll out each; and spread over them thinly cut slices of streaky bacon, sprinkle well with chopped parsley, roll up into 2 rolls, and fasten the edges and ends carefully. Tie up each in a cloth and boil for about 2½ hours. Take up, remove the cloth, cut the pudding into slices, dish up, and serve hot.

Dinner No. 19. For 6 persons. Total cost, 3s.

Vermicelli Soup	9d.
Roast Mutton & Baked Potatoes	1s. 6d.
Savoury Cabbage	4d.
Cornflour Blancmange, with Marmalade (see p. 57)	5d.
	3s. 0d.

Vermicelli Soup

3 lb. meat bones, 1 ½ oz. dripping, a small bunch of parsley or other herbs, 2 or 3 slices bacon, 2d. worth of soup vegetables, 3 quarts water, 1 oz. vermicelli.

Prepare the vegetables, and insert 2 cloves in the onion. Break up the bones, and fry them till brown in a stew-pan with the fat ; next add the water with a little salt, bring to the boil, and remove the scum; put in the prepared vegetables, the herbs and bacon, and again bring to the boil, skim ; and continue to simmer for 1 ½ or 2 hours. Strain the broth, boil up, add the vermicelli, and cook slowly for about 10 minutes; then season to taste with salt and pepper. The soup is then ready to serve.

Savoury Cabbage

Remove the outside withered leaves from 2 small or 1 large cabbage, then cut the cabbage into quarters lengthways and cut out the stalk portion. Wash thoroughly in cold water and allow to soak for about 15 minutes in salted water. Next put the cabbage in sufficient boiling water with a very small piece of soda, and cook quickly in a saucepan for about an hour, when the cabbage should be quite tender. Drain the cabbage in a colander and chop it finely. Melt an oz. of dripping in the saucepan and fry in it to a golden colour ½ a peeled and finely minced onion, stir in a large tablespoonful of flour, then add the chopped cabbage and a little stock, milk or water. Stir till it boils, and let it cook gently for another 10 minutes. Season to taste with more salt if needed, also a little pepper and a grate of nutmeg. The dish is then ready to serve.

Dinner No. 20. For 6 persons. Total cost, 3s. 6d.

French Bread Soup	11d.
Beef or Rabbit Pie (see p. 28)	1s. 6d.
Parsnip Croquettes	9d.
Coffee Mould (see p. 57)	4d.
	3s. 6d.

French Bread Soup

Take sufficient stale bread to weigh about 8 or 10 oz., cut it into dice with or without crust, but it is best if the crust is first rasped off with a grater (the raspings can be used as breadcrumbs). Put the cut pieces of bread into a stew-pan (a casserole for preference), add to it a dessertspoonful of salt, a teaspoonful of sugar, and a ½ teaspoonful of pepper; pour over about a quart of warm water, and let it stand covered for 20 minutes. Peel and mince finely an onion, and fry it in an oz. of dripping to a pale golden colour (not brown) ; to this add 2 or 3 slices of bacon, or an equal quantity of ham or gammon cut into strips or chopped up small, and fry also for a few minutes. Next put both the onion and bacon into the pan containing the bread, etc., add also about a pint of milk, and allow to cook gently for about an hour. Stir the soup frequently during the process of cooking, using a whisk for preference, as this will help to make the bread smooth. Before serving see that the soup is correctly seasoned; a grate of nutmeg will improve it. A small leek, trimmed, washed and cut up very small, may be used instead of the onion mentioned in the recipe. When meat or bone stock is available, use it in place of water; the soup will then be all the richer. Should the soup be found too thick, a little more milk or water may be added.

Parsnip Croquettes

Wash and peel thinly 3 large parsnips, cut them into slices, and chop them finely, or else pass the slices twice through a mincing machine.

Mince finely 3 rashers of bacon, and fry them in a saucepan with an oz. of dripping, then put in the minced parsnips, and stir over the fire for a few minutes. Sprinkle over about 2 tablespoonfuls of flour or double the quantity of breadcrumbs, or failing these use soaked bread free from crust, mashed up into a pulp. Mix well and moisten with a little stock or water, season with salt, pepper, and a grate of nutmeg. Let the whole cook gently for about, 1 an hour, or until the parsnips are quite tender. Pour off any moisture, and spread the parsnip mince on a large plate or dish to get cold. Make it up into even-sized portions, and shape each on a floured board into croquettes (cork or ball shapes). Beat up an egg with a little milk or water; with it brush over each croquette, next re-shape, then roll in breadcrumbs. Have ready some hot frying fat, and fry the croquettes to a golden brown. Take up when done, drain and dish up. Serve hot.

Chapter 13

Some Economy Hints

Substitutes for Meat and Fish

Cheese, peas, beans, lentils, oatmeal, nuts, may be substituted with advantage at all times in place of meat and fish.

Avoid Waste in Food Scraps

Make use of all food scraps. Clean peelings of many vegetables help to make a basis for vegetable stock. Clean fruit parings and corings may be stewed in water with sugar, strained, and used as a flavouring medium. Bacon rinds will help to flavour meat stocks. Stale cheese can be grated, and either eaten raw, with some combination of starch food, or made into savoury dishes. Odds and ends of fat should be melted down, clarified, and used for dripping. Remains of cold vegetables may be reheated in a frying-pan with a little fat and seasoning. Stale bread can be put to a multitude of uses; as raspings (after baking) in puddings, for lightening purposes, in savoury stuffings for meat and fish. Cold meat and cold fish may be converted into curry, savoury mince, pasties, rissoles, pies, patties, shapes, etc. Skim milk takes the place of wholemilk if a little extra fat is substituted from some other source. Take care of bones. They should make at least two successive courses of good soup, with the addition of vegetables, cereals, and pieces of bacon. Kernels of plums, prunes, greengages, and other stone fruits, should be kept, milled, and used for flavouring. They much improve the taste of all jams. Sour milk, with a little carbonate of soda, makes an excellent raising agent for scones, etc.

Avoid Waste in Cooking

Adopt long, slow cooking methods in preference to quick, galloping methods. Cook as much as possible in a steamer, or in a double saucepan, as this preserves the flavour of the foods better. When stewing, or boiling, use very little liquid, and serve the food, if possible, with or in this liquid. Put your pots on early in the morning, and keep up only a moderate fire until dinner time. This saves the gas or coal, the pocket, as well as the flavour of the food. It is well to remember that if rice, tapioca, sago, barley, etc., are soaked in cold water or milk for some time before cooking, they will require less time to stew or bake tender. This again is a saving of fire.

Eat Plenty of Vegetables

Potatoes are cheap, and very wholesome. They should be cooked in their skins, so that none of the goodness is lost. The Irish people largely live on potatoes, with an occasional piece of fat bacon, and the Irish peasant is a man of great strength. Carrots, onions, tomatoes, leeks, parsnips, turnips, cabbage, etc., should be steamed or stewed rather than boiled in a lot of water. Dried vegetables must be well washed and soaked for some time before using.

EAT PLENTY OF FRUIT

Bananas and oranges are cheap and plentiful just now; the former are especially nutritive. White grapes are also good and plentiful. Every season has its own natural supply of fruits, if we only take the trouble to look for them. Dried fruits must be washed, and soaked all night before using. Fruit acids cleanse the blood, and help to make bone, brain, and muscle. Remember the saying, "An apple a day keeps the doctor away."

DEAR SUGAR!

Never mind! Eat less or don't eat it at all! Use treacle, golden syrup, honey, currants, sultanas, figs, raisins, dates, etc. instead. Any of these will supply the body with much more sugar than is contained in the grocery article, and they have the advantage of possessing this sugar in a ready state of assimilation.

DEAR BUTTER!

Never mind! Don't eat it! Use dripping, or try fats made from nuts or pure olive oil. Eat plenty of nuts. Peanuts are only 5d. per lb., and contain 50 per cent. of oil in their composition. Grind some daily, or if you don't want the trouble of this, pay an extra penny per lb. and get it done for you.

Dear Fresh Fish!

Never mind! Don't eat it if dear! Try the dried varieties instead. Kippers, bloaters, and salt or pickled herrings all contain more nourishment than fresh fish, and at the present moment are only about one half the price.

Meat Is Steadily Rising

Never mind! Eat less of the prime, expensive joints. There are plenty of cheap cuts still available. Sheep's head (7d.) makes an economical meat dish and a soup. The water in which the head is boiled will furnish an excellent basis for Scotch Broth. Brisket of beef (5d. and 6d.) makes an excellent stew. Boiled meat puddings, composed of "beef trimmings," or "butchers' pieces," are by no means to be despised. "Beef skirtings" serve quite as well as expensive steak for a pie or pudding. Shin of beef (6d.) is both nourishing and tasty, as is also breast of mutton (6d.). Oxtail (7d.) makes an appetising stew. A sheep's heart (5d.) stuffed is a savoury and economical dinner for two people. Tripe (6d.) is a dish often despised by the poor, and relished by the rich. It is exceedingly economical and nutritious, and deserves to find its way to everybody's table. Half a pig's head (5d. per lb.) is particularly nice when really well cooked. The "cheek" will make a dinner dish, while the coarser parts can be turned into collared head, an excellent breakfast dish.

The House-Keeping Money Won't Last!

Make your family eat more slowly, chew the food well, and masticate more thoroughly. You will find that in this way less food is needed, for it goes much further.

The 3d. Tax Makes Tea Almost Prohibitive

Then drink less of it, or have something else. There are plenty of other beverages quite as good. Try cocoa; it is far more nourishing than tea. Or coffee and chicory, if you can't afford the real article. Barley or oatmeal water, flavoured with lemon rind, makes a refreshing and nourishing drink, and is especially good for children of all ages.

Tinned Fruits

Fresh foods are better than tinned foods, but when it is necessary to use the latter, the following precautions should be observed. Buy only the best brands by well-known makers. Reject tins that bulge on the surface in any way. Don't keep any food in the tin after it has once been opened. Empty out the contents, and use them up as quickly as possible.

Wholemeal Flour

This is more nourishing than white flour, and an excellent corrective of constipation. Wholemeal flour is specially useful in the making of bread, cakes, and scones.

EQUIVALENT WEIGHTS AND MEASURES

FOR DRY MATERIALS.

16 oz. = 1 lb. = 3 Teacupfuls or 2 Breakfastcupfuls, closely filled but not heaped.
8 oz. = ½ lb. = 1 Breakfastcupful, closely filled but not heaped.
4 oz. = ¼ lb. = 1 Teacupful, loosely filled.
1 oz. = 2 Tablespoonfuls, filled level.
½ oz. = 1 do. do.
¼ oz. = 1 Dessertspoonful, do.

FOR LIQUIDS.

4 Gills = 1 Pint = about 3½ Teacupfuls, or nearly 2 Breakfastcupfuls.
1 Gill = 1 Small Teacupful.

APPROXIMATE WEIGHTS IN COINAGE.

2 oz. = 10/- in Silver, or Six Penny Pieces.
1 oz. = 5/- in Silver, or Three Penny Pieces.
½ oz. = 2/6 in Silver, or a Penny Piece and a Halfpenny Piece.
¼ oz. = 1/3 in Silver, or a Threepenny Piece and a Halfpenny Piece.

INDEX

A

B

C

W

Y